Felipe de Souza Miranda
Gilberto Petraconi
Alexei Essiptchouk

Coatings deposited by Solution Plasma Spray

Felipe de Souza Miranda
Gilberto Petraconi
Alexei Essiptchouk

Coatings deposited by Solution Plasma Spray

Nanostructured and compositionally graded materials applied in the aerospace sector

ScienciaScripts

This book is a translation from the original published under ISBN 978-613-9-61001-3.

Publisher:
Sciencia Scripts
is a trademark of
Dodo Books Indian Ocean Ltd. and OmniScriptum S.R.L publishing group

120 High Road, East Finchley, London, N2 9ED, United Kingdom
Str. Armeneasca 28/1, office 1, Chisinau MD-2012, Republic of Moldova, Europe
Printed at: see last page
ISBN: 978-620-7-50386-5

Table of contents:

Coatings deposited by Solution Plasma Spray

Nanostructured and compositionally graded materials applied in the aerospace sector

AUTHORS: Felipe de Souza Miranda Gilberto Petraconi Filho Alexei Essiptchouk

Thanks

A greater challenge than developing the project and writing the thesis is to be able to thank everyone who has helped and supported me during these four years of my doctorate.

First of all, I would like to thank the most important person in my life, my mother Mathilde, who has always shown the strength, determination and perseverance to face all the difficulties that life has thrown at her, serving as an inspiration and example for my life. Thank you for always supporting me, encouraging me and giving me the strength to never give up on my dreams. Mom, if I could choose a thousand times, I would choose you!

To my brother Fernando for showing that there are no limits that life can impose on us. You've always looked after me, my brother and best friend, we've often quarrelled, but we've always sought the best for our lives, I love you!

Elias, the father I was able to choose and who accepted me as his son. All these years you've been at the base of our family, fighting for us, you're an example. Thank you very much.

To Aline, my friend, girlfriend and wife, for always supporting and understanding me throughout the years we've been together. Always with positive words to keep me from getting discouraged. I love you every day and every day much more!

To Felipe Caliari, my friend and battle buddy, who has helped me through all these years of my doctorate without once refusing to do so; with my work and yours, we were able to develop this system. You will achieve all your goals because you are one of the most determined people I know. Thank you my friend!

To Professors Dr. Gilberto Petraconi and Dr. Alexei Essiptchouk for accepting me as a student and for giving me all the guidance and support I needed to carry out this project.

To Professors Dr. Gilmar Patrocínio Thim, Dr. Argemiro Soares Sobrinho and Dr. Douglas Leite for making their laboratories available, which were so essential for carrying out my experiments.

To Professor Homero Santiago Maciel for his advice and for being the first to trust in my potential.

To my friend Marco Antonio de Souza, you were to blame for all this!

To my friend Cristian for the relaxed conversations, advice and exchange of knowledge, which contributed greatly to the completion of the project.

To my colleagues in the lab for the conversations and friendship built up over the years. Especially Tiago Moreira for helping me characterize and discuss the results of the coated samples.

To my friends from the physics workshop, Cláudio Garufe and Jorge Prado, for always being willing to help with the construction of our system. Without their enormous skill and experience in machining and design, it would not have been possible to complete this work.

To all the professors and staff of the Postgraduate Program in Aeronautical and Mechanical Engineering.

I would like to thank ITA for the opportunity and CNPq for the financial support throughout my doctorate.

And finally, to everyone who contributed in some way to making this research possible.

Summary

Carbon/Carbon (C/C) composites are widely used in thermo-structural components, especially in the aerospace sector. However, carbon has low resistance in ablative environments, limiting its application in thermal protection systems. This problem can be overcome by protecting the composite with passivation layers based on ceramic materials with low catalytic efficiency in reactions with oxygen and nitrogen at high temperatures. In this context, the main objective of this work is the synthesis of thick coatings with a compositional gradient between SiO_2/SiC deposited on C/C composites using a new process called *High Velocity Solution Plasma Spray* (HVSPS). These coatings are formed from reactions between the C/C composite and the $Si(OH)_4$ colloidal solution sprayed by a supersonic plasma jet generated by a direct current torch. The plasma torch has a different design from conventional torches, increasing the interaction time with plasma regions at high temperatures. The relatively rapid cooling of the particles sprayed by the plasma jet on impact with the substrate contributes to the formation of an amorphous coating, as well as to the synthesis of nanostructured materials. The formation of the compositional gradient between SiO_2/SiC in relation to the thickness of the coatings was investigated as a function of the exposure time to the plasma jet. The formation of SiC with a predominant concentration at the coating/substrate interface acts as a bonding layer between the substrate and the SiO_2 layer predominant on the surface of the coating exposed to the plasma jet, reducing the incompatibility between the thermal expansion coefficients, as well as showing a process of chemical adhesion of the coating in reactions between the Si sprayed by the plasma jet and the C in the composite. The outer layer of SiO_2 in the coating has the ability to fill or seal cracks (defects) in order to protect the composite. Thermal tests, using an oxyacetylene torch (ASTM E285-80 standard), show a reduction in the area of cracks present in the coatings, as well as the establishment of a crystallization process of the SiC atomic structure. On a theoretical level, the behavior of the particle's velocity and temperature along its trajectory in interaction with high enthalpy and velocity plasma jets was evaluated, considering solid particles with a defined diameter.

Chapter 1

1.   Introduction

Carbon fiber reinforced carbon matrix composites (C/C) are widely used materials in structural components, especially when exposed to intense aerothermodynamic loads (PALANINATHAN, 2010). Its high strength and exceptional fracture toughness, combined with its refractory properties, low density, high resistance to erosion, corrosion and wear make this material ideal for use in structural components subjected to high temperatures, such as turbines and atmospheric re-entry vehicles. When used in inert atmospheres or in a vacuum, C/C composites maintain their properties at temperatures above 2000°C (WEBSTER, 1996). However, in oxidizing environments at high temperatures, carbon fibre-based composites suffer intense degradation due to the high catalyticity of reactions between carbon and oxygen, making it difficult and often impossible to use them in aerospace devices (KIM, 2005). This problem can be overcome by depositing environmental barrier coatings to protect the C/C composite against oxidation, loss of mass and consequently its mechanical properties (SAVAGE, 1993).

Coatings that protect the substrate against thermal oxidation, failure due to high temperatures or thermal flow are classified as Environmental Barrier Coatings (EBR) (LEE, K. N. U., 2000; ANDO, 2005; ENGEL, 2014). There are basic principles for selecting a material to be used as an environmental barrier coating, such as: withstanding reactive environments, having low oxygen permeability (so that it is not transported to the substrate) and, finally, being chemically compatible with the substrate (LEE, K. N. U., 2000). For effective protection, it is desirable to deposit or form oxides in the coating, such as Al_2O_3, SiO_2, ZrO_2 and others.

The oxide layer minimizes the diffusion of carbon out of the substrate and prevents oxygen from entering its interior (WEBSTER, 1996). Materials with self-sealing capacity, i.e. the ability to fill or seal 17 cracks (defects) can effectively protect the substrate. This class of material is known as *self-healing materials*; in this sense, vitreous or glass transition materials are extremely suitable for sealing cracks or defects in the coating (WOOL, 2008a).

Silica (SiO_2), being a *self-healing* oxide with low oxygen permeability, can be used to protect C/C composite materials. Generally, C/C composites are coated with materials capable of consuming part of the incident thermal energy. This energy is used during the oxidation reactions that take place, for example, in the re-entry atmosphere. For this reason, coatings containing silicon are widely used (PALANINATHAN, 2005). The oxidation of silicon-based ceramics at high temperatures typically generates a uniform, dense SiO_2 film with low oxygen permeability, providing oxidation protection for C/C composites (KAISER, 2008).

Therefore, a lot of research is being carried out to incorporate SiO_2 or SiC into ceramic coatings. Various techniques are used for depositing coatings, including chemical vapor phase deposition (CVD), carburizing and plasma spray (LIU, 2012). Among these, the plasma spray technique has high potential for application due to its flexibility in processing a wide variety of solid or liquid materials, making it possible to deposit on substrates with complex geometries and achieving high deposition rates

(FAUCHAIS, 2004). In the process of plasma spray deposition from colloidal solutions, a liquid precursor containing the main element (or elements) of the desired coating composition is used. The coating can be formed in reaction with the working gas of the plasma torch, between the components of the solution, with the deposition atmosphere and even with the substrate used. In addition to these characteristics, this process makes it possible to obtain coatings with nanometric scale graphics, correlating with lower defect indices, greater tolerance to deformation and better porosity distributions, particularly in ceramic materials (XIE, 2004; FAUCHAIS, 2008a, 2008b). In addition, the reactivity of plasma can be exploited by combining the plasma spray process with material synthesis in a single step, making it possible to obtain materials with a chemical/compositional gradient (FORD, 1999; MI, 2017). Reactive plasma spray processes can produce composite coatings by reacting two or more materials (DAI, 2017).

In this context, the main focus of this work is the development and adaptation of a plasma torch for processing liquid precursors, aiming at a complete system for the synthesis of thermo-structural coatings applied to thermal protection systems. The main result of the work refers not only to the development of the plasma thermal spray deposition system, but also to the process of synthesizing nanostructured coatings with a compositional gradient of different materials that act as an efficient passivation layer on the C/C composite. Specifically, graded SiO_2/SiC coatings were obtained, formed in the plasma/precursor and precursor/substrate interaction process. The formation of SiC at atmospheric pressure (not reported in the literature for this type of process) is due to the characteristics of the process developed in this work called *High Velocity Solution Plasma Spray* (HVSPS). This process allows the efficient deposition of a coating with a compositional gradient of SiO_2/SiC without the need for an interlayer, preventing problems of void formation, since the coating is deposited in a single step. The studies of the microstructural properties of the coatings, as well as the evaluation of their self-healing capacity, are presented as the results of the development and improvement of the supersonic plasma jet thermal spray system combined with the use of colloidal precursor solutions of silicon oxide for the synthesis of passivation coatings on carbon/carbon composites.

Chapter 2

2.   Main objective

The main objective of this work is to develop a supersonic plasma spray system for processing liquid precursors with a view to synthesizing environmental barrier coatings with a compositional gradient between nanostructured SiO_2/SiC on C/C composites.

2.1.   Specific objectives

In order to achieve the main objective of this work, it was necessary to define some specific objectives, which are described below:

- Adapt and modify the plasma torch to operate with liquid precursors;
- Develop a precursor solution injection system with flow control;
- Developing a deposition process using a silicon oxide precursor solution on C/C composites;
- Define the effective operating characteristics of the plasma torch, such as current, voltage and power, taking into account the parameters of plasma spray deposition processes;
- Obtaining environmental barrier coatings with a compositional gradient between SiO_2/SiC that act as a passivation layer on C/C composites in oxidizing environments;
- Characterize the chemical and microstructural structure of the coatings obtained;
- To develop a numerical method for theoretically evaluating the particle's velocity and temperature along its trajectory and when interacting with high enthalpy and velocity plasma jets, considering solid particles with a defined diameter.

Chapter 3

3.   Literature review

3.1. Object of study

Carbon in its graphitic form remains structurally stable at temperatures of up to 3000°C in non-oxidizing atmospheres in the sense that its mechanical and thermal properties are not altered when compared to those of various other ceramic or metallic materials. Only after 2500°C is there a slight plastic deformation (BUCKLEY, 1993). These properties have a direct bearing on the use of carbon-based thermo-structural composites in thermal protection systems. Obviously, in order to maintain these properties, the composites depend directly on the quality and type of carbon fibers, the orientation of the carbon fibers, the precursor matrix, the heat treatment temperature, the microstructure of the matrix and the manufacturing techniques.

Although C/C composites are produced with carbon fibers that have low strength on their own, the final material has a higher tensile strength than superalloys and ceramics at temperatures above 1000°C (BUCKLEY, 1993), as shown in Figure 1.

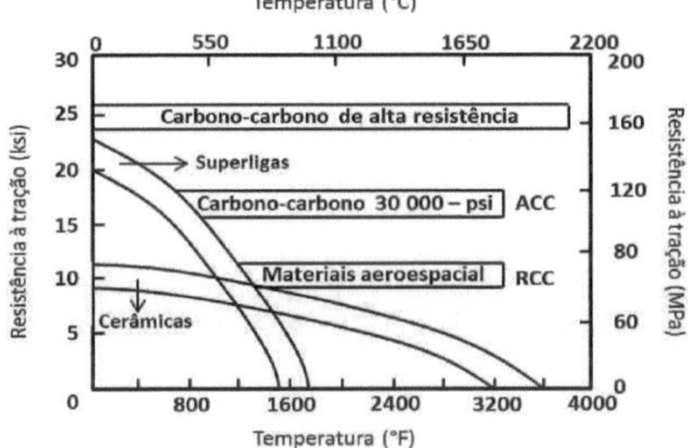

Figure 1. Comparison between the tensile strength of regular (RCC) and advanced (ACC) composite materials, superalloys and ceramics (adapted from (BUCKLEY, 1993)).

Because of these characteristics, C/C composites are widely used in re-entry vehicles, which require materials that are not only low in density but also capable of withstanding extreme thermomechanical conditions. In this application, C/C composites are mainly used in the leading edges and stagnation points of aerospace devices. These regions are the most affected as they receive the greatest thermal and mechanical loads during flight.

The limited use of C/C composite materials is due to carbon being a highly reactive element, especially with oxygen. In aerospace applications, this reactivity is the main problem to be overcome, making it necessary to find solutions that minimize or even eliminate the loss of its thermomechanical properties when subjected to oxidizing environments.

To minimize the degrading effect of oxidation, hybridization of the matrix with silicon carbide (SiC) alloy has been a widely used alternative (SILVA, 2011). Although SiC composites reinforced with carbon fibers have greater stability in oxidizing environments, this still cannot be considered a permanent solution, since it only slows down the oxidation of the carbon fibers (CHENG et al., 2000).

3.1.1. Catalyticity of a material

The erosion rates of different materials depend not only on the oxidation regimes, but also on the phenomenon of catalyticity. A catalytic surface can considerably increase the temperature of the sample by means of additional heat flow in the material and thus shift reactions towards the active oxidation regime (FAUCHAIS, 2014). Knowledge of the catalytic efficiencies of materials used in environmental protection systems is particularly important when analyzing the effects of cracks in coatings. The heat fluxes and surface temperature values reached by space vehicles during the re-entry phase on a planet will depend on the emissivity and catalytic efficiency of the materials used.

thermal protection materials used in their construction. To this end, materials with high emissivity values and low catalytic efficiencies must be obtained. Catalyticity can be defined as the catalytic efficiency of a material in relation to the recombination (on the surface) of atomic species generated in the environment due to the chemical reactivity of unsaturated valence atoms on the surface.

A vehicle re-entering a planet at hypersonic speed produces shock waves, which in turn generate excited species (ions, atoms, molecules, electrons) that diffuse into the boundary layer and react with the materials covering the vehicle. Atomic oxygen and nitrogen, present during the re-entry phase on Earth, collide with the surface of the vehicle and recombine to form volatile molecular species (O_2, N_2, NO) through exothermic reactions causing the surface temperature to rise, leading to an active oxidation mechanism (SILVA, 2011).

Equations (1), (2) and (3) exemplify a catalytic reaction with nitrogen and atomic oxygen reacting catalytically with Si(g) (SILVA, 2009):

$$Si_{(g)} + N \rightarrow SiN \tag{1}$$

$$SiN + N \rightarrow Si_{(g)} + N_2 \tag{2}$$

$$Si_{(s)} + O_2 \rightarrow SiO_2 \tag{3}$$

Thus, during the development phase of a protective material, the contribution due to the recombination of atomic species on the surface must be taken into account in order to predict heating rates on the exposed parts of the vehicle. These atomic recombination reactions are described by heterogeneous catalysis models and usually depend on the molecular structure and morphology of the surface (LEVY, 2006).

The catalytic efficiency of the material in the atomic oxygen recombination process can be evaluated by means of a direct method for measuring the recombination coefficient γO, defined as the ratio between the flow of oxygen that recombines on the surface and the flow of atomic oxygen that reaches the surface of the sample (KOVALEV, 2005). Measurements carried out on the C-SiC composite in a reactive air plasma atmosphere at 200 Pa total pressure show that this material has low values for the yO coefficient, however, its value increases exponentially with the temperature of the sample surface, starting at a value of approximately 0.001 at 800 K and reaching values of approximately 0.070 at 1800 K (ALFANO, 2009a). The atomic oxygen recombination coefficient (yO) can be obtained by determining the atomic oxygen concentration gradient in the vicinity of the sample surface using actiometry and optical emission spectroscopy techniques (ALFANO, 2009b).

Another way of estimating catalytic efficiencies of materials is by evaluating the surface temperature of the sample. Due to the identical testing conditions that samples are exposed to, a direct comparison is possible between the different radiated heat fluxes. Using this method, Dabalá et al (DABALÁ, 1995) evaluated the catalyticity of various materials (SiC, Si, C, C/SiC and SiO_2) in a reactive plasma environment of a mixture of argon and oxygen at low operating pressures (5 to 10 mbar). Within the experimental errors involved, it was shown that the radiated heat flux calculations of Si or SiO_2 are approximately equal to those of SiC, which indicates that the catalytic efficiency of these materials is practically equivalent.

In this scenario, it can be concluded that in a composite that uses carbon fiber as a reinforcement and has no coating (passivation layer), when exposed to an air plasma environment at high temperatures, the exothermic reactions of recombination of atoms on the surface of the sample with atomic oxygen and nitrogen are intensified. In other words, the surface of the composite is more catalytic than that of the passivation layer material, with the main consequence of obtaining higher saturation temperatures on the surface of the sample, leading to higher erosion rates.

Therefore, depositing surface layers (coatings) is considered an efficient alternative to actually protect the composite, reducing the catalytic efficiency of the composite for exothermic reactions with atomic oxygen and nitrogen. For this reason, various studies have been carried out to determine compositions and ways of obtaining passivation coatings on carbon-based composites used in environmental protection systems.

3.2. Coatings: an overview

Coatings are applied to substrates (metallic, ceramic, polymeric or composite) to incorporate into their surface one or more characteristics or qualities that they did not originally possess, such as increasing the useful life of a material/equipment, increasing its resistance to corrosion, wear, oxidation, thermal protection or even providing a gain in efficiency, depending on its application (COUNCIL, 1996).

Increasing the performance of a given component can mean, in addition to greater longevity of use, a reduction in costs related to design and subsequent maintenance. In this context, the choice of material is just as important as the design of the final product. New materials are being developed for the most diverse applications, which are subjected to extreme conditions, such as high temperatures, abrasive wear, oxidizing and corrosive atmospheres or even a combination of these conditions. It would be impossible for a single material to meet all these requirements, even for special alloys. This contributes to the increase in demand for environmental barrier coatings that can be deposited on substrates made of a wide variety of materials (FAUCHAIS, 2014) and with properties tailored to specific applications.

Coatings can be divided into thin and thick films. Thin films have thicknesses of up to 20 μm and are typically synthesized by chemical vapor deposition (CVD) or physical vapor deposition (PVD). However, most thin film deposition processes require reactors working at low pressures, increasing process costs and limiting the size and geometry of the substrate to be coated.

Another category of coatings are thick films, which have thicknesses greater than 20 mm or even a few millimeters. These are applied when the protection performance depends directly on the thickness of the coating, for example in environmental barrier coatings for aerospace devices, which are subjected to severe erosion, corrosion and oxidation environments. Thick film deposition methods include chemical/electrochemical plating, brazing, solder overlays and plasma spray (FAUCHAIS, 2014).

3.2.1.   Environmental barrier coatings

The large number of industrial processes that operate in highly aggressive environments are characterized by high temperature, high temperature gradients, high pressure, high stress on the elements that make up the system, particulate materials and the presence of oxidizing and corrosive atmospheres. Components used in industry, such as steam turbines, gas turbines, coal conversion, oil refining and nuclear power generation have high temperature and oxidation as the main failure mechanisms. In the aerospace sector, for example, fuel is injected into turbines and combustion causes an increase in pressure and temperature. In the exhaust nozzle, temperatures can reach 1650°C, and during cooling they are in contact with the atmosphere, making them susceptible to oxidation (WRIGHT, 1991).

In order to help protect these systems, various studies have been carried out into the deposition of surface layers, the primary objective of which is to prevent the base material from being subjected directly to the aggressive environment and losing its specific desired properties (LEE, K. N. U., 2000). To protect composite or metallic materials used in such extreme conditions, environmental protection coatings or environmental barrier coatings (RBA) are applied. The main requirements for these coatings are (GHOSH, 2015):

- low specific weight,
- low thermal conductivity,
- provide reactivity with the oxidizing environment, preventing it from occurring with the base material (if applicable),
- withstand varying stresses related to heating and cooling and resist thermal shock waves.
- deformation capable of minimizing the effects of thermal expansion,
- radiate heat.

Figure 2 shows a schematic of the main function of oxidation protection layers, also known as passivation layers. The low oxygen permeability of the coating isolates the substrate from the oxidizing environment and prevents carbon from diffusing out of the substrate (WEBSTER, 1996).

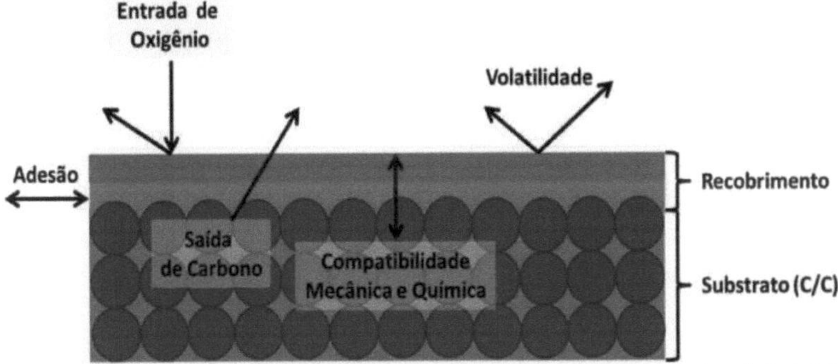

Figure 2: Factors to consider when building an integrated oxidation protection system (adapted from (WEBSTER et al., 1996)).

The first coatings to protect against oxidation and corrosion were typically aluminum-based, with the aim of forming aluminum oxide in their final composition. Single-layer coatings composed of oxides are more resistant to oxidation because they have a lower oxygen permeability.

Mullite ($Al_2O_3SiO_2$) was a much-studied material in the 1980s because of its compatibility with composite materials and also because of its chemical stability. Multilayer coatings were also deposited using high melting point oxides (ZrO_2, YO_3, etc.), but these coatings were very fragile and susceptible to failure through cracking (SARKISOV et al., 2008).

Recently, a variety of materials have been applied as coatings to meet the needs of industry (TKACHENKO et al., 2012), some of which will be listed below:

-The carbides SiC, TiC, TaC, ZrC and HfC.

-The borides ZrB_2, HfB2 and TiB_2.

-The oxides SiO_2, TiO_2, ZrO_2, B2O3 and HfO_2.

-The nitrides Si3N4 and BN

   -Combined groups: SiC + TaC, $Si_3N_4 + SiC$, ZrC + SiC, HfC + SiO_2, $ZrB_2 +$ SiC.

Due to its excellent mechanical and thermal properties, ZrB2-based coatings have been widely investigated for the protection of C/C composites. However, studies report that during the oxidation of ZrB2, ZrO2 and $B_2O_3 \text{ are formed}$. At 1000°C the B2O3 is in the gas phase and only the very porous and discontinuous ZrO_2 remains in the coating, causing the protection system to fail (FAHRENHOLTZ, 2005; HU et al., 2009). These effects can be minimized by adding silicon-based materials such as MOSi2 and SiC to zirconium diboride. These materials help to form borosilicate (SiO $B_{22}O_3$) and zirconium silicate ($ZrSiO_4$) which have low oxygen permeability, preventing the material from being directly exposed to the oxidizing environment (KAISER et al., 2008; LI et al., 2013).

Tului et al (TULUI et al., 2010) proposed coating composites with a high concentration of SiC (up to 66%) and ZrB_2. The gain in oxidation resistance is due to the reactions that occur between the SiC and ZrB_2 of the coating and the oxygen in the atmosphere, resulting in the formation of SiO2 and ZrO2 which act as a passivation layer preventing the substrate from oxidizing and maintaining its mechanical properties at temperatures above 1600°C.

Resistance to oxidation also increases when a SiC interlayer is added between the substrate and ZrB_2 coatings, increasing the interfacial distance between the materials. The addition of SiC promotes the formation of SiO_2 in the coating, which decreases oxygen permeability (HWANG et al., 2007). In most processes for coating ceramic materials on carbon/carbon composites, it is necessary to synthesize a *bond coating* that provides good adhesion between the substrate material and the coating material. SiC is the material most commonly used for this purpose as it is highly compatible physically and chemically with C/C composites. The aim of depositing an interlayer is to reduce the appearance of cracks in the coating, as they have compatible thermal expansion coefficients. Often voids are formed between the interlayer and the coating, which reduces the stability and protection of the substrate (HUANG et al., 2006). In order to overcome these challenges, many studies, such as those carried out in this thesis, are looking for deposition techniques that synthesize high-density nanostructured coatings by single-layer formation processes with a compositional gradient, combining the main qualities and making up for the shortcomings between the materials that make up the coating.

- Coatings with a compositional gradient

Adhesion strength, thermal fatigue or oxidation are directly related to the materials that make up the coating and the substrate. One way of reducing or eliminating the influence of these factors on protective performance is to use coatings with a compositional gradient of different materials (KOKINI et al., 1997). The advantage lies in the development of a coating with specific properties from more than one material. Coatings with a compositional gradient can be continuous (a single deposition) or in layers, always respecting the compatibility between the materials. However, depositing coatings in layers can lead to the formation of voids, reducing, for example, adhesion to the substrate in ablative environments (KIEBACK et al., 2003) such as those generated in rocket nozzles and in atmospheric entry conditions for aerospace devices (BELMONTE et al., 2009).

Coatings with a compositional gradient can be obtained by various deposition techniques, such as CVD, PVD or PECVD thin film deposition. However, films synthesized by these techniques are not usually used in RBA's, mainly due to the requirement for thick coatings over large areas.

Thermal spray processes such as *High Velocity Oxygen* Fuel (HVOF), Flame Spray or Plasma Spray processes are used for depositing thick coatings (DAVIS, 2004). For plasma spray processes, the system must contain two powder feeders or use two simultaneous plasma torches. However, the deposition parameters are not easily controlled in relation to the materials used (size, density, morphology and fluidity) and the plasma torch may not reach the temperature required to melt both materials (KHOR et al., 1999). One way of overcoming these obstacles is to use precursor materials with a predefined composition, which can be previous mixtures of materials with close melting points or even liquid suspensions or solutions.

Schulz et.al (SCHULZ et al., 2003) carried out a study comparing the oxidation resistance of multilayer coatings and coatings with a compositional gradient in a single layer. The results obtained by applying laser thermal shock in an oxidizing environment demonstrate the relative superiority of single-layer graded coatings.

Figure 3 shows the results of the tests between the conventional coating and the compositional gradient coating, comparing the number of cyclic loads (fatigue test) and the maximum temperature withstood in an oxidizing environment.

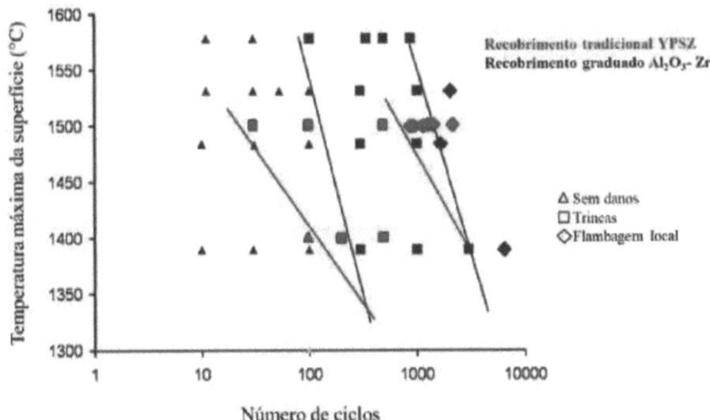

Figure 3: Comparison between conventional and compositionally graded coatings (adapted from SCHULZ et al., 2003)

The conventional coating (red lines) of Zirconia partially stabilized with Yttria (YPSZ) withstood lower temperatures (~1500°C) when compared to the coating with compositional graduation (black lines) between Al2O3 -Zr (~1600°C). The Al2O3 , present in the compositionally graded coating, which is an oxide, has lower oxygen permeability, helps protect against oxidation, since the YPSZ coating was consumed to form ZrO_2. However, the biggest difference observed was in relation to thermal fatigue wear between the two coatings. At temperatures above 1000°C, the conventional coating showed the appearance of cracks which increased as the test progressed. On the other hand, the gradient coating only begins to show cracks with a greater number of cycles, resisting temperatures greater than 1500°C.

In addition to the use of compositional variants, the addition of materials with the ability to seal cracks that may appear during the application of the coating has also been studied. This ability increases the useful life of a protection system or structural components, making them less susceptible to failure. Materials that possess this ability are classified as *self-healing* materials (ZWAAG, VAN DER, 2010).

- .2.1.2 Self-healing coatings

Self-healing materials are polymers, metals, ceramics and their composites which, when subjected to some kind of damage through thermal or mechanical shock, have the ability to self-heal and restore their original properties (WOOL, 2008b). *Self-healing* is considered one of the most important abilities of a material, referring to overcoming problems such as decreased integrity (cracks) due to damage caused during application. *Self-healing* can occur automatically, i.e. the moment the material suffers damage that compromises its integrity, maintaining its working conditions without any compromise to the system. When choosing to use a material with this functionality, possible system failures must of course be analyzed in accordance with the requirements of the application. In the case of structural components and coatings, the main causes of failure are surface cracks, fatigue, thermal shock, oxidation and corrosion (GHOSH, 2008).

Self-healing cracks in coatings can be achieved by adding oxides or oxide precursor materials, with Si-based materials being the most widely used in environmental protection systems. Crack sealing by oxidation was first reported by Lange and Radford (LANGE, F.F. AND RADFORD, 1970). In this study, polycrystalline SiC samples were heat-treated and compared with untreated samples and then subjected to oxidation tests to check the material's ability to seal cracks. The results show that the treated samples had an increase in sealing efficiency of 10% more than the untreated ones, evaluated in relation to the area of the cracks. However, the most important result of these tests was the observation that the cracks were filled by oxide materials formed during the tests as a result of the SiC reacting with oxygen. Crack sealing is initiated at high temperatures, above 1000°C, for Si-based materials, but depends on the test conditions (GHOSH, 2008). This sealing process can be seen schematically in Figure 4, where oxygen penetrates the interior of the substrate reacting at high temperature with the SiC, forming the silicon dioxide (SiO_2) sealant.

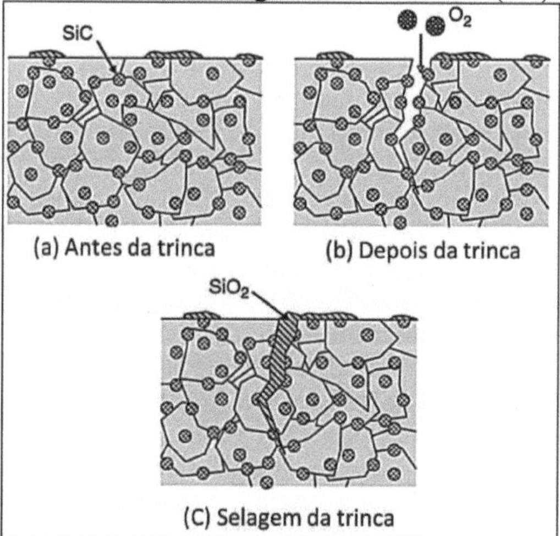

Figure 4 - Schematic illustration of crack *self-healing* (adapted from (GHOSH, 2008)).

3.3. Plasma spray deposition process

3.3.1. Plasma torches: basic concepts

Plasma torches are devices used to stabilize an electrical discharge with a gas flow, in order to convert electrical energy into thermal energy. In a thermal plasma torch, operating from an electric discharge, high enthalpy plasma results from the interaction of the gas with the electric arc. The study of electrical discharges in gases and the formation of the plasma jet involves phenomena of gas dynamics, mass and heat transfer, electrophysical and aerothermodynamic processes (SOLONENKO, 2003).

Plasma torches can be classified according to the source of electrical energy, i.e. electrical arcs can be generated from a direct current (DC) or alternating current (AC) source, or by the type of discharge used, which can be transferred arc or non-transferred arc. Transferred arc torches have one of the electrodes external to the torch body, through which the arc extends from the internal electrode. Due to the transport of electric current in the plasma jet generated, this configuration forms high enthalpy plasma jets compared to non-transferred arc torches. For non-transferred arc torches, both electrodes are positioned inside the torch. This means that the electric arc remains confined to the discharge channel and the plasma jet generated does not carry an electric current (ZHUKOV et al., 2007).

When designing a thermal plasma torch, in addition to the type of electrical energy source, consideration must also be given to the enthalpy and temperature of the plasma jet suitable for the application, the choice of appropriate materials, the implementation of a stabilization system and control of the arc length (if applicable). In the case of non-transferred arc torches, the electric arc can be stabilized by vortexing the gas that forms the plasma. In most cases, the method of self-fixing the arc length by the *shunting* effect is used. In addition, a magnetic field generated by one or more solenoids can be used to fix the arc length. In this case, the radial part of the arc moves axially to the place where the magnetic field is greatest. Furthermore, the interaction between the magnetic field and the electric arc current produces the driving force that moves the arc tangentially, avoiding the positioning of the arc spot at a fixed point, thus reducing electrode erosion (wear) (ESSIPTCHOUK et al., 2009).

The materials used in plasma torches, especially those exposed to the electric arc, are subjected to a high thermal gradient and erosion at the location of the electric arc. Properties such as specific heat, melting temperature, thermal expansion coefficient, thermal conductivity, work function and electrical resistivity must be considered when choosing these materials (CALIARI, 2016).

Figure 6 shows a generic schematic of a non-transferred arc plasma torch with the electrodes (cathode, 1, and anode, 2) arranged concentrically and between which the electric arc is established (MARQUESI, 2016). The gas vortex produced by the vortex chamber, 4, installed between the insulated electrodes (3), is used to stabilize the arc. To fix the arc at the anode, a magnetic field produced by the solenoids, 7, is applied. The electrodes are subjected to high heat flows and to maintain their functionality, they are cooled by a continuous flow of water through the cooling jacket, 6.

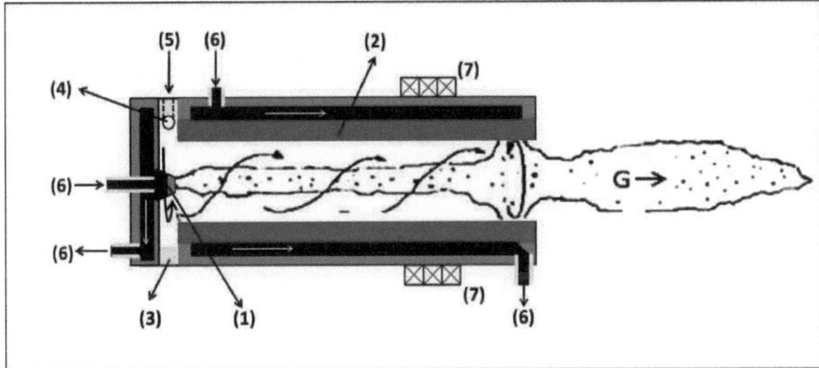

Figure 5 Schematic of a linear thermal plasma torch, (1) hot cathode, (2) anode, (3) electrical insulator, (4)

vortex chamber, (5) gas inlet, (6) cooling water inlet and outlet, (7) solenoid (adapted from (SILVA, 2011)).

Although there is a great diversity of plasma torch constructions, their principle of operation is based on the generation of plasma flow due to the forged convective interaction of a gas with the electric arc established between two electrodes (ZHUKOV et al., 2007). The application of a relatively high potential difference, which oscillates at a high frequency, transforms the electrically insulating gas into a conductor and forms an electrically conductive channel. The high temperature resulting from this channel reduces the electrical resistance of the medium due to the increase in the number of charged particles, thus allowing a high current to pass through the gas, establishing the arc. The electrons, accelerated by the electric field in the region between the electrodes, transfer their kinetic energy to the heavy particles through collisions, raising the temperature of the gas, dissociating its molecules, exciting and ionizing the atoms, factors which contribute to increasing the degree of ionization of the plasma. The electric field generated near the cathode accelerates the charged particles, which collide with the cathode surface. As the mobility of heavy particles is much lower than that of electrons, an excess of positive space charge forms in the region near the cathode. This phenomenon increases the electric field in the vicinity of the electrode which, in turn, facilitates the emission of electrons from the electrode (due to the tunneling effect) intensifying the electron density in the plasma (ZHUKOV et al., 2007). The constriction of the arc at the cathode (another important phenomenon) increases the current density and, respectively, the thermal flux at the cathode surface, increasing its temperature and the emission of electrons due to the thermionic effect.

3.3.2.   Plasma spray

Plasma spraying emerged after the Second World War as a surface finishing technology. Today it is widely used for depositing thick coatings (from hundreds of micrometers to a few millimeters) on a substrate in order to protect it in an aggressive environment or improve its function. It is commonly used in many industrial sectors, including aeronautics, energy, automotive, mining, biomedical and electronics (TUCKER et al., 2013). The synthesis of coatings using the plasma spray deposition technique occurs through the stacking of lamellae resulting from the impact, flattening and solidification by the collision of molten particles (DAVIS, 2004). The precursor material for the coating can be in the form of powders, wires, molten materials, solutions or suspensions (HERMANEK, 2001). What distinguishes the plasma spray process from other technologies is its applicability to a wide variety of materials, including metallic and refractory materials. In a plasma spray system, the most commonly used plasma torches are non-transferred arc torches, as shown in Figure 6.

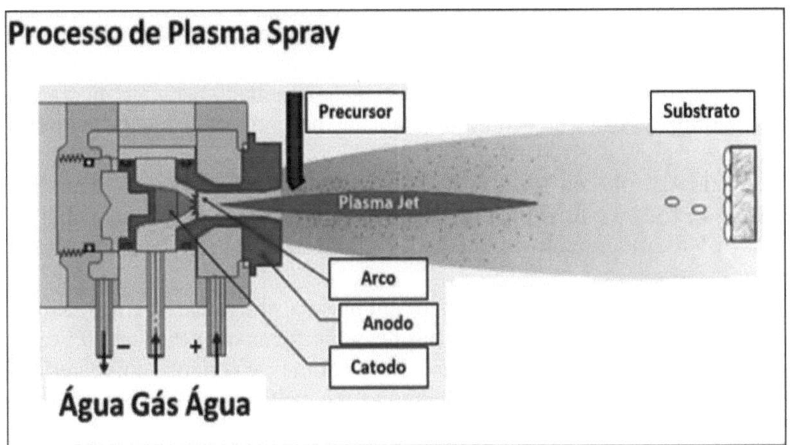

Figure 6. Schematic drawing of a non-transferred arc plasma torch used in a deposition process (adapted from (HUB, 2014)).

In addition to the plasma torch, a plasma spray system basically consists of:

I. Electrical power source, gas supply system, precursor supply (powders or solutions) and associated controls;

II. Equipment for preparing the material for coating, such as those for injecting powders by charging gas into the plasma region to melt the particle and for controlling particle size distribution and morphology;

III. Chamber with controlled atmosphere if necessary (with humidity control) and operation also at low pressure;

IV. Mechanical equipment for controlling the relative movement between the plasma torch and the substrate.

In the plasma spray technique, a carrier gas drives the material particles by injecting them at high speed through the plasma, where they are melted or partially melted, taking the form of droplets that are deposited and solidify on the surface to be coated. The material to be deposited is taken in the form of a solution or powder to a torch with sufficient enthalpy to melt it (FAUCHAIS, 2004; XIE, 2004). The parameters of the plasma spray process, as well as the characteristics of the precursor (solids or liquids) used for coating, influence the properties of the deposited materials. For example, the high temperatures of the plasma jet and the high cooling rates of the material during the coating process can also promote the formation of amorphous phases (SAMPATH, 2001; CAO, 2004). Coating characteristics such as porosity, atomic structure, roughness, cohesion and adhesion are fundamentally related to the effect of the precursor's interaction with the plasma jet (GUO, 2005; XIE, 2006).

The main driving force for the manufacture of thick coatings by plasma spray is their high deposition rate. A few kilograms per hour of raw material can be processed with torches with a power level of a few tens of kilowatts at a relatively low operating cost.

Plasma spraying is probably the most versatile of all thermal spraying processes, because there are few limitations on the materials that can be sprayed, and few limitations on the material, size and shape of the substrate (FAUCHAIS, 2004). The coatings are characterized by a highly anisotropic lamellar structure. In addition, the stacking of the particles generates specific inter-lamellar characteristics along the structure, mainly voids, which may or may not be connected by the particles that subsequently encounter the coating.

Conventional plasma spray (CPS) processes use powders with a particle size ranging from 10 to 100 iim. Typically, these result in coatings formed by lamellae with a thickness of the micrometre order and a diameter of a few tens to hundreds of micrometres.

The interest in developing and studying plasma spray coatings with nanometric rather than micrometric characteristics has been the focus over the last 30 years. This interest stems from the improved properties of nanometric coatings compared to those of micrometric size (GELL, 1995). Reducing the structure to the nanometric scale produces an improvement in the hardness, reduction of defects (voids), modulus of elasticity and thermal conductivity of the coating. One of the main drawbacks of processing nanometric particles by plasma spray is the difficulty in injecting them into the core of the high enthalpy jet. In this case, it is necessary to adjust the angle of injection and the flow of gas transporting the material so that it is not so intense that it crosses the plasma jet or so mild that it cannot reach the center of the jet (FAUCHAIS; ETCHART-SALAS; et al., 2008). These drawbacks, associated with the injection of very small particles into the plasma jet, can be overcome using the following techniques (PAWLOWSKI, 2008):

- Using a suspension with nanometer-sized particles, with the carrier gas replaced by a liquid injected in the form of a jet or atomized. This technique is called *Suspension Plasma Spray* (SPS).
- Using precursors in solutions that form nanometric particles on the fly, called the *Solution Plasma Spray Process* (SPSP). This technique significantly limits the safety issues associated with handling nanometric particles and avoids many of the drawbacks associated with suspension stabilization, especially when different materials (e.g. metal alloys and oxides) are mixed.

3.3.3. Plasma spray with liquid precursors

Since the 1990s, nanostructured materials have been considered as a new concept for increasing the performance of engineering components. Many studies have been carried out on the properties of nanostructured materials used in structural components and coatings. Ceramic materials have gained prominence mainly because they have greater hardness due to their smaller grain size, greater wear resistance and lower incidence of defects (FAUCHAIS, 2015; GELL, 1995). Obtaining nanostructured coatings by plasma spray processes is only possible using liquid precursors. In the case of solid precursors (powders) with a nanometric distribution, there is a great difficulty in relation to fluidity in the feed line to the plasma torch, causing intermittent injection of the material. The fluidity problem is overcome by increasing the flow of the carrier gas, but on the other hand, such an increase can cause the axis of the plasma jet to shift, producing a non-uniform coating on the substrate. Another factor is that the nanoparticle may not penetrate the center of the plasma jet, inhibiting fusion processes and accelerating towards the substrate to form the coating

(XIE, 2004; DELBOS, 2006; VISWANATHAN, 2006; GELL, 2008; PAWLOWSKI, 2009).

The *Solution Plasma Spray* technique allows thick nanostructured coatings to be deposited without the need for a very sophisticated infrastructure. Expenditure on precursor materials is relatively much lower than when using powders. The flexibility of using different raw materials makes it possible to explore a variety of precursor liquid compositions, adjusting their concentration according to the desired application. It is also possible to exploit the reactivity of the plasma to obtain a final coating with a different composition to the original precursor (MIRANDA, 2017).

To form the coating using liquid precursors, these must be injected into the plasma region in smaller droplets (pulverized or atomized). In

The solvent is then evaporated, forming the solid material which is melted (or forms a shell) and accelerated towards the substrate, as illustrated in Figure 7.

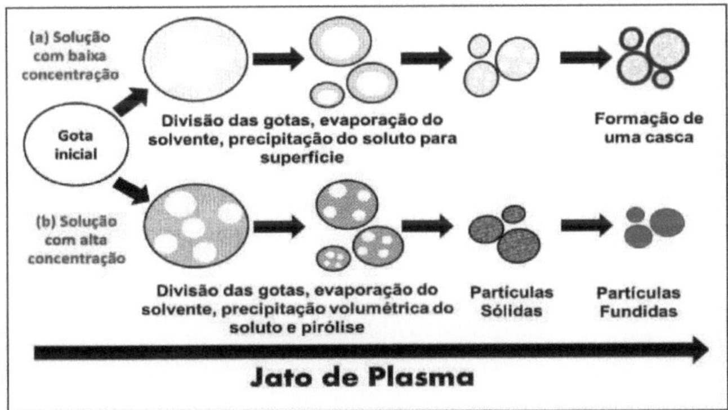

Figure 7. Formation of the material from a liquid precursor with low and high concentration of the material to be deposited (adapted from (FAN, 2016)).

All these steps take place almost instantaneously, and it is essential that the plasma generator supplies the right energy to process the liquid precursor and form the material.

3.3.4. Liquid precursor injection methods

The method of solution injection directly influences the plasma spray deposition process, due to the cooling of the plasma caused by heat exchange with the liquid (FAUCHAIS, 2014). Liquid precursors can be injected in two ways: by atomization or by mechanical injection. As shown in Figure 8, the atomization process consists of injecting a liquid at low speed through a nozzle, where this liquid is fragmented by a gas (usually argon due to its high specific mass) (PAWLOWSKI, 2009). The geometry of the nozzle, the viscosity and density of the liquid, as well as the amount of gas used, determine the size of the droplets. The liquid can be charged using a pump or a pressurized vessel. This method generates small disturbances in the plasma jet because it has similar trajectory behavior to a gas.

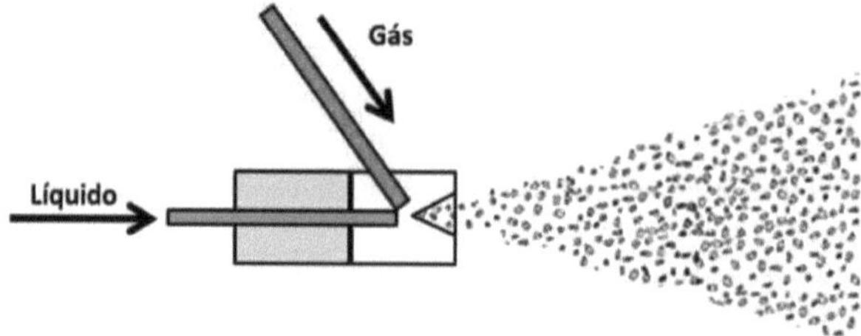

Figure 8. Schematic representation of the atomized injection method

The mechanical injection process consists of injecting a liquid into a pressurized container through a nozzle, as shown in Figure 9. As the liquid is injected in the form of a concentrated jet, this method directly influences the operational stability of the plasma torch in the case of axial injection through the discharge channel.

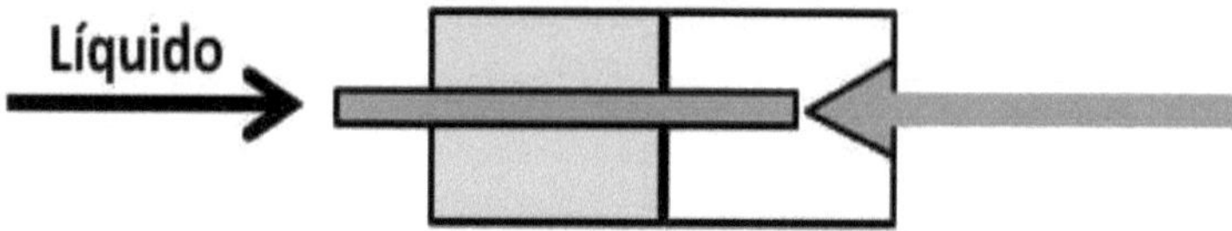

Figure 9. Schematic representation of the mechanical injection method.

Chapter 4

4.   Materials and methods

In this work, a plasma thermal spray system was developed for processing liquid precursors with a view to synthesizing passivation coatings on C/C composites.

The materials used and the experimental procedures carried out during the research are described below.

4.1. Experimental system

As shown in Figure 10, the experimental apparatus used in the deposition process consists of the following subsystems:

- Tandem plasma torch,
- Electric power source and an ignitor connected in series,
- Cooling system,
- Gas supply system,
- Operational control and data collection system,
- Sample holder with sample rotation and temperature control,
- Precursor material injection and dosing system.

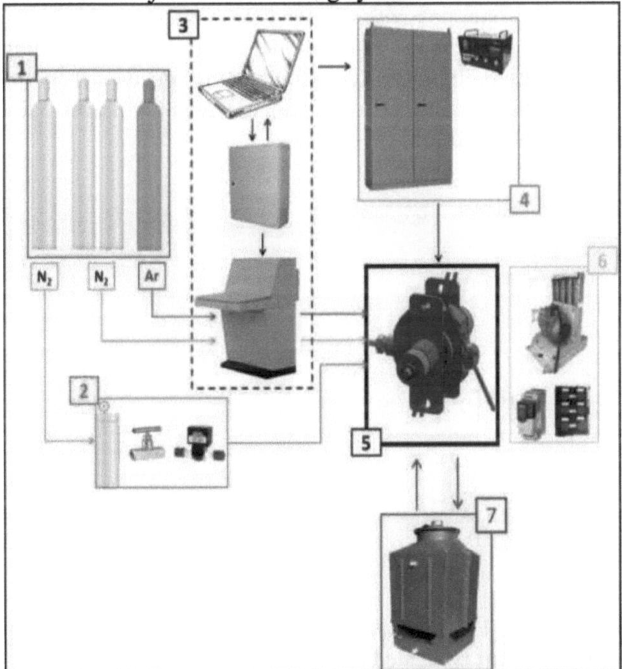

Figure 10. Experimental apparatus: (1) Gas supply system, (2) Material injection and dosing system , (3) Operating control and data collection system, (4) Power source and igniter connected in series, (5) Tandem plasma torch, (6) Sample holder, (7) Cooling system.

Figure 11 shows the layout of the main components of the deposition system.

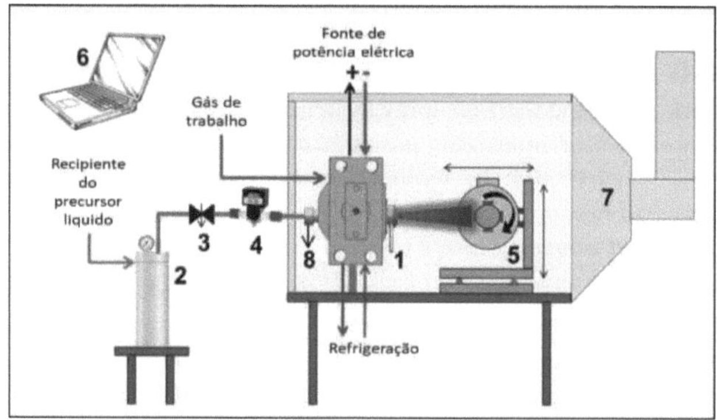

Figure 11. Complete injection system where (1) plasma torch, (2) reservoir, (3) needle valve, (4) flow meter, (5) sample holder, (6) control panel, (7) exhaust system and (8) spray nozzle.

A detailed description of the main subsystems is presented below.

4.2. Tandem plasma torch

The thermal plasma torch (Figure 12) used in this project is of the non-transferred arc type and operates with nitrogen flow rates (working gas) of between 150-450 l/min. According to design calculations, the torch generates plasma jets with temperatures in the 2000 to 6000 K range at supersonic speeds of 1200-1600 m/s. The torch operates with currents ranging from 55 to 135 A and voltages between 240 and 360 V under the process conditions investigated.

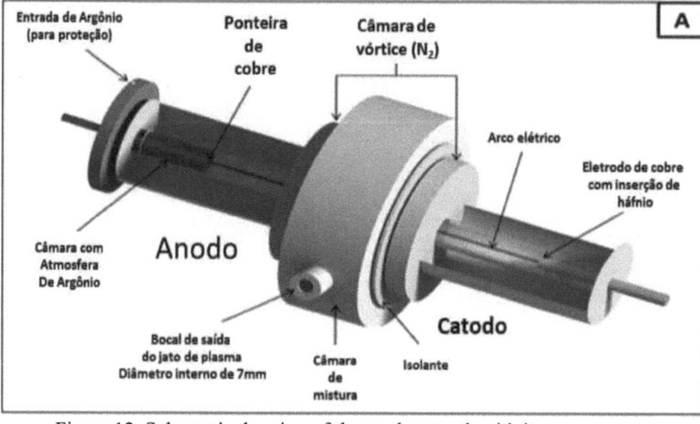

Figure 12. Schematic drawing of the tandem torch with its components.

As shown in Figure 12, the general assembly of the plasma torch is divided into 3 main parts: the central part is located between the electrodes; the other two parts refer to the electrode regions (cathode and anode) of the plasma torch.

Figure 13 shows an image of the plasma torch in operation during the deposition process.

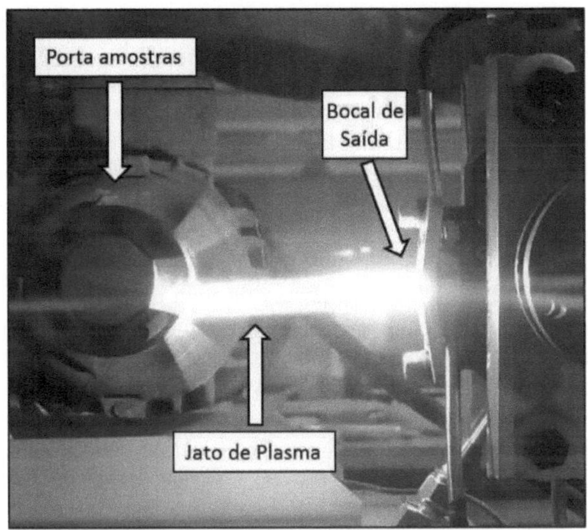

Figure 13. Image of the plasma torch in operation during the deposition process.

4.3. Electric power source and ignitor

Figure 14 shows a functional schematic of the DC power source with reactive control. The source can supply up to 70 kW of power to the plasma torch. Basically, the source consists of a three-phase isolating transformer, a passive rectifier bridge (with two diodes per phase), three variable reactors, an independent direct current source (called the control DC source), a variac and a filter to block RF coming from the ignition.

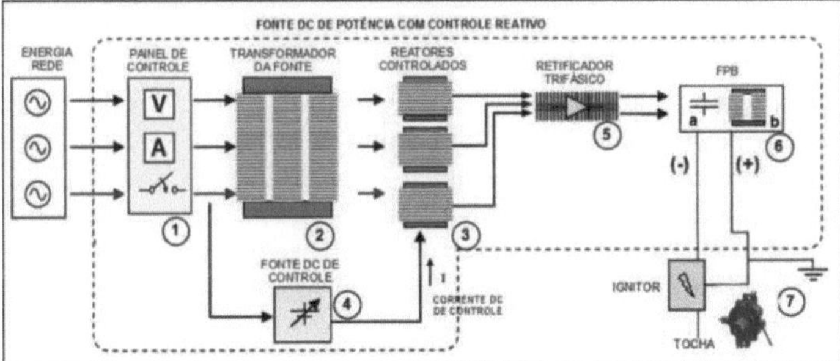

Figure 14. Schematic of the DC Power Source with Reactive Control

In the diagram in figure 14, the power from the three-phase network is fed to the source via a control panel (1). The panel is made up of contactors, electrical protection elements, multimeters and control for activating or interrupting the source. The output of the panel feeds the source transformer (2) and the current control (DC) source (4). The transformer is used to adapt the mains voltage to the voltage and current levels required to operate the plasma torch, depending on the application. Each of the transformer's three-phase lines is connected to an independent reactor, represented by coils (3). These reactors have the function of limiting the current supplied by means of their inductive reactance. This reactance is controlled by a DC current supplied by the control (DC) source (4). The higher the current supplied by the control source, the lower the inductance of the reactors and the more current is released to the three-phase rectifier (5). The control source acts on the 3 reactors in series, so the same DC current controls the three reactors, maintaining the balance between the phases.

The three-phase rectifier is an uncontrolled type (with diodes). The output of the rectifier is direct current, with a minimum ripple of 4% which increases according to the current applied. In tests, maximum values of 7% were obtained for the torch's maximum permitted operating current of 150 A. To protect the source, a low-pass filter FPB (6) was installed. The DC current passes through an FPB, whose function is to isolate the source from high-frequency peaks from ignition and rapid arc oscillations.

As for the operational aspect of the source, the rectifier used in the source already provides a typical ripple of 4%, which is enough not to affect the stability of the electric arc in the plasma torch. The power control is carried out by an independent current source, with currents from 0 to 25 A adjusted by a variac, with the maximum power corresponding to 3% of the main source power. A variac adjusts the control current of the independent source, which acts to saturate the magnetic core of the variable reactors, releasing power from the main source. Once the voltage between the electrodes of the plasma torch has been established, the arc is ignited by activating a high-frequency igniter with high-voltage pulses of around 7kV. This produces a pilot discharge responsible for forming a conductive channel in which the electric arc develops between the electrodes.

4.4. Refrigeration and gas supply system

The plasma torch cooling system used in the tests consists of a cooling tower, a heat exchanger and a water reservoir, all interconnected to a closed circulating circuit driven by a hydraulic pump.

The water flow rate is adjusted manually, controlling the closed circuit pressure up to its maximum value of 10 kgf/cm^2 , limited by the characteristics of the hydraulic pump. Water flow is monitored using an Omega model FTB 691A digital meter, with an operating range of between 3.8 and 38 L/min.

To supply the torch and its dependent subsystems with working gas, a system was built comprising four lines: two of Nitrogen (N_2), one of Argon (Ar) and a reserve line. N2 is used as the working gas and Argon is used as the anode protection gas. Argon is also used to facilitate the torch ignition process by reducing the electrical resistivity of the gas in the discharge channel.

Figure 15 shows a flowchart of the gas supply and control system, consisting of a panel equipped with pressure gauges (inlet and outlet), flow regulators, fast-opening electromechanical (solenoid) valves (used mainly in the ignition process) and flow meters.

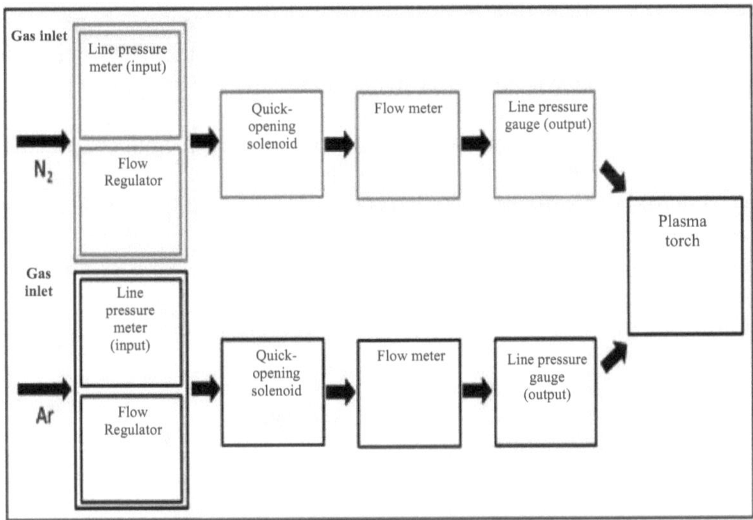

Figure 15. Flowchart of the gas supply and control system

The specifications of the gas control system components are shown in table 1.

Table 1. Components of the gas control system.

Component	Make/Model	Operating range	Error(%)
Flow Controller	VSI/EMO-85M-24	0 - 500 (l/min)	-
Flow meter	SMC/PF2A551	50 - 500 (l/min)	± 5
Pressure gauge	Rücken / RTP-420	0 - 10 (bar)	± 0,2
Solenoid (Open/Close)	Werk-Schott/2W200-20	40 (ms)	-

All the components of the gas supply system panel have digital outputs for data acquisition, which allows automatic monitoring of gas flow and pressure.

4.5. Operational control and data acquisition system

To facilitate the operation of the deposition system, an automated operational control and data acquisition system was designed and built. The system consists of a Micrologix 1400 micro-CLP with sixteen analog channels,
thirty-two digital input and output channels and six thermocouple input channels. The control interface (Figure 16) was developed in "ladder" language, using Rockwell Automation's "micrologix" software. The interface allows the following functions to be carried out:

- Monitoring the temperature of the samples,
- Monitoring the operating pressure of the plasma torch,
- Monitoring the pressure and water flow of the cooling system,
- Monitoring and control of working gas flow (N_2),
- Argon (Ar) flow monitoring and control,
- Monitoring and control of power source parameters,
- Monitoring the operation of systems.

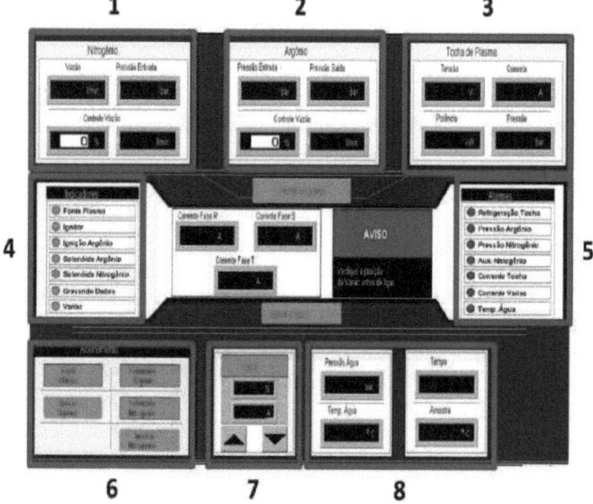

Figure 16. Command and monitoring interface of the experimental system, where: (1) Control and monitoring of the working gas, (2) Control and monitoring of the anode protection gas, (3) monitoring of the electrical parameters of the plasma torch, (4) Indicators of correct system operation, (5) System fault alarms, (6) Gas activation buttons, (7) Control of the operating current and (8) Monitoring of pressure, cooling water temperature, experiment time and sample surface temperature.

4.6. Sample holder

An essential procedure for plasma spray deposition systems is the adjustment of the operating conditions for the interaction between the plasma jet and the substrate, influenced by the preheating temperature of the sample, the exposure time to the plasma jet and the cooling rate. These factors can be controlled by independent movement between the samples and the plasma torch or even with simultaneous movement between the two (dynamic deposition), and therefore directly influence the properties of the coatings obtained. If it is impossible to move the plasma torch (as in the present study), it is necessary to build a sample holder that can perform the movement, and some requirements must be met for its construction:

- Adjusting the distance of the sample(s) from the plasma torch (influences the structure, deposition rate of the coatings and temperature of the samples);
- As many samples as possible being processed with the same experimental parameters, i.e. the same torch operating conditions (enthalpy), deposition material flow rate and, in the case of the solution, the same manufacturing conditions;
- Preheat monitoring (depending on the substrate material, a preheat temperature setting is required to guarantee its integrity and also the adhesion of the coating);
- Monitoring the temperature of the samples during the process;
- Control of sample rotation speed.

All the requirements described above were taken into account when designing and manufacturing the sample holder. Figure 17 shows the sample holder built with the capacity to process 8 samples simultaneously. The system also has rotation control via a frequency inverter connected to a three-phase motor with 1471 Watts of power (Figure 17(C)). The semi-automatic three-axis positioning control (Figure 17(D)) is used to center the samples in relation to the plasma jet and adjust the deposition distance.

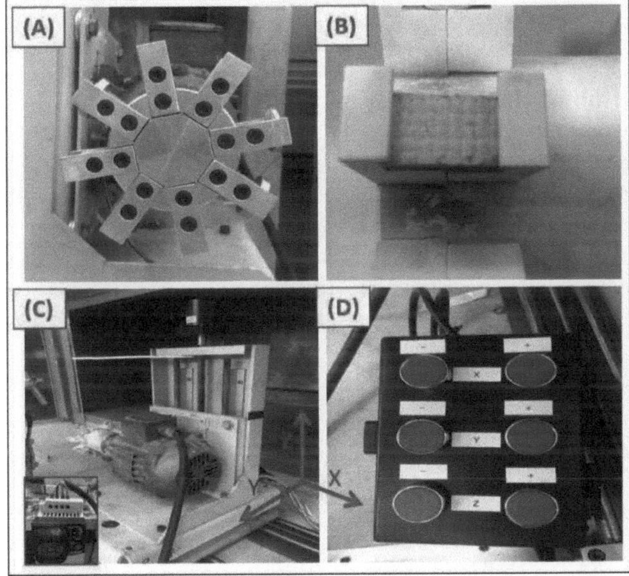

Figure 17: (A) Sample holder with capacity to process 8 samples simultaneously, (B) Sample fixture, (C) Motor, sample positioning system in three directions (X, Y and Z) and highlighting the frequency inverter for rotation control and (D) Sample positioning control.

The temperature of the samples is monitored during the experiments using a RayTek infrared pyrometer in the range -20 °C to 500 °C. The pyrometer has a digital output for real-time data acquisition via the control and data collection system described in section 5.5 (Operational control and data acquisition system). The position of the pyrometer sensor, shown in Figure 18, must respect the maximum distance of 12 cm from the surface on which the temperature measurement will be taken. The pyrometer was positioned and fixed using an aluminum rod and the distance from the samples was adjusted directly on the positioning table on its Z axis.

Figure 18. Position of the pyrometer in the experiment in relation to the samples.

4.7 Preparation of the C/C samples

In this work, the thermo-structural Carbon/Carbon composite developed by the IAE/DCTA materials group was used as a substrate.

The carbon fiber composite samples with phenolic resin were obtained from carbon fiber fabric prepregs. The carbon fiber used to make the samples was bidirectional Plain Weave (Plain Fiber). Table 2 shows the characteristics of the carbon fiber according to the manufacturer's specifications.

Table 2. Carbon fiber properties according to manufacturer's specifications

Properties	Values
Weight	196 g/m2
Filaments/cable	3000
Thickness	0.22 mm

The polymer used as the impregnating material was resol-type phenolic resin CR-2830, supplied by Crios Resinas Sintéticas S/A. The prepregs had a phenolic resin content of 50% by mass. After the molding process in a hydraulic press, the composites molded with carbon fibers had a mass concentration of 58% reinforcement fibers and 42% resin.

The 20x10x4 (mm) parallelepiped-shaped samples were sanded on the edges to reduce irregularities caused during cutting. Figure 19 shows the samples fixed to the sample holder.

Figure 19. Detail of how the samples were attached to the sample holder (left figure) and the completed sample holder (right figure).

4.8 Preparation of silicon oxide precursor solutions

In this section, the procedures for obtaining the silicon oxide precursor solution (Silanol - (Si (OH)$_4$)) are described. In the procedure, the ion exchange resin must first be regenerated by adding a 2 mol/L solution of nitric acid (HNO3) in sufficient quantity to cover it. After a period of 24 hours, the resin is washed with distilled water until the pH of the filtrate reaches the pH of the wash water (pH 6), measured with indicator paper. A qualitative test for the presence of sodium is then carried out (flame test). If the test result is negative, the resin is fit for use. After this stage, sodium metasilicate (Na2SiO^5H2O) in aqueous solution (10% m/m) is passed through an ion exchange resin (IR120 - Rohm and Haas), thus obtaining the silicon oxide precursor solution (Silanol - (Si (OH)$_4$)). The procedure for obtaining the liquid precursor is shown more succinctly in Figure 20 by means of a block diagram.

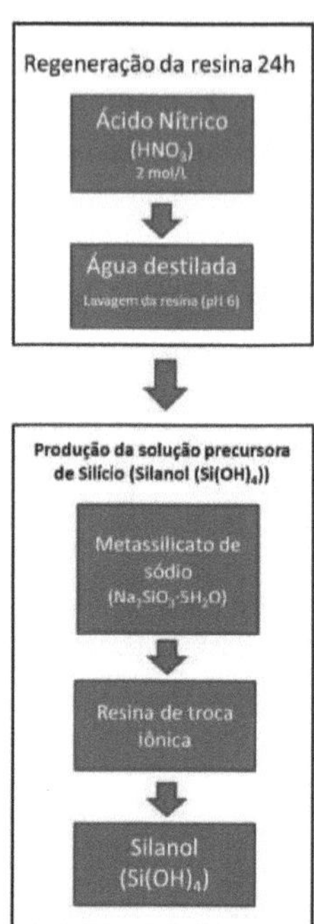

' Figure 20. Block diagram of the production of the Silicon Oxide precursor solution

1.9. Material injection and dosing system

The liquid precursor injection system developed in this study was designed and built with a view to using both the atomized injection method and the mechanical injection method, as shown in Figure 21, since the use of liquid precursors for deposition in a tandem plasma generator has not yet been reported in previous studies.

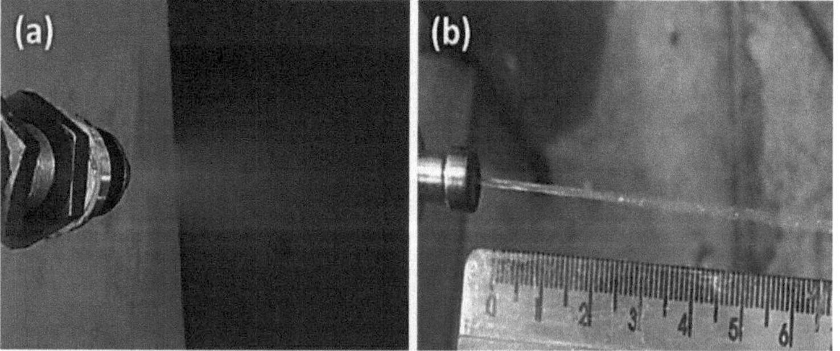

Figure 21. Injector nozzle of the deposition system: (a) atomized injection and (b) mechanical injection.

Injection tests were carried out to verify the influence of the material injection method on the operation of the plasma torch, as well as to obtain the characteristics of the injection system, such as: material dosage control, operation of the drag method (by pressurized container), atomizing nozzle diameter and mechanical injection. The results of the tests indicated that in the atomization process the torch operates under better conditions of voltage and current stability, as well as inhibiting the occurrence of nozzle clogging. The liquid precursor injection system is shown schematically in Figure 22.

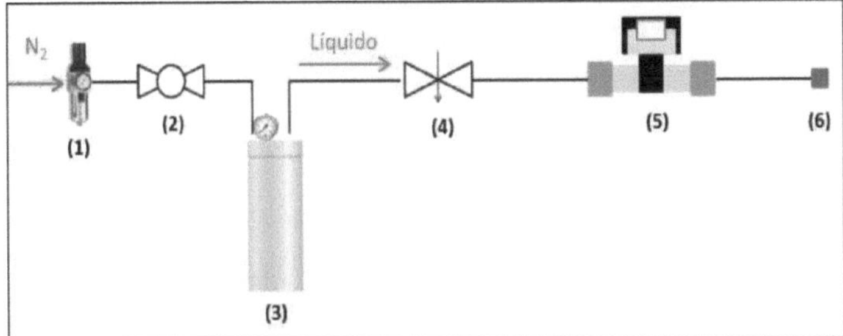

Figure 22. Liquid injection system, where: (1) Pressure regulator, (2) Ball valve, (3) Reservoir, (4) Needle valve, (5) Flow meter and (6) Spray nozzle.

The liquid precursor atomization and injection system consists of a pressurizing gas inlet with a pressure regulator (1), a ball valve (2) to release the gas passage for pressurizing the reservoir (3) with a storage capacity of up to 1 liter, equipped with a pressure gauge for monitoring pressure. For fine control of the passage of the liquid precursor, a needle valve (4) was added to the line in series with a flow meter (5) for liquids, with a measuring capacity of 30 to 300 ml/min and finally an atomizing nozzle (6) coupled to the plasma torch.

A good design of the spray nozzle responsible for injecting the precursor liquid into the torch is essential both for maintaining the proper operating conditions of the plasma torch and the conditions of the deposition process. Figure 23 shows the schematic drawing of the nozzle used in the liquid precursor deposition experiments.

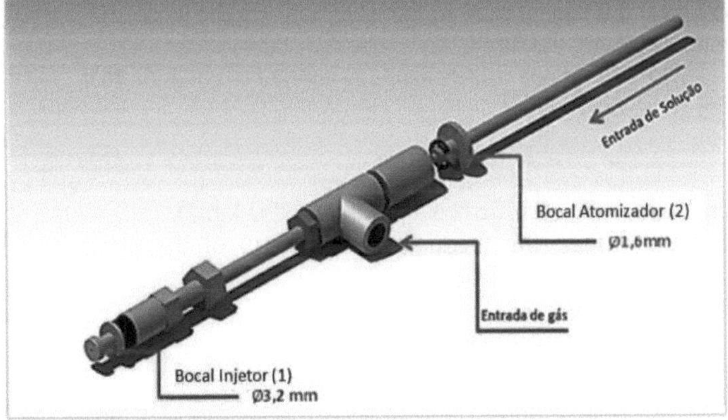

Figure 23. Spray nozzle with component details

The injector nozzle (1) has a 3.2 mm diameter outlet. Another important aspect of this nozzle is that the atomization is carried out before it reaches the spray nozzle, by the atomizing nozzle (2) in order to avoid or minimize the deposition of material in the spray nozzle and consequent clogging.

1.10. Tandem plasma torch energy characterization

The energy characterization of the plasma torch is an important procedure for analyzing its operating conditions in terms of its thermal efficiency (conversion of electrical energy into thermal energy) and average enthalpy of the jet, with a view to the proper processing of materials. For the characterization of the torch, the values of:

- Working gas flow (G_{N2});
- Current (I);
- Voltage (U);
- Total power (P_{total});
- Cooling water flow (G_{H2O});
- Water temperature variation between the torch inlet and outlet (T).

In order to estimate the average enthalpy ( ) of the plasma jet, we first need to The total power (P_{total}) applied to the arc discharge is calculated from equation (4).

- **Total power (Ptotal):**

$$P_{total} = U.I \tag{4}$$

Next, the dissipated power (P_{AB}) of the plasma torch is calculated. The values are obtained from the water flow rate (G_{H2O}) of the cooling system, the specific heat of water ($cpH_2 o$) and the temperature variation (ΔT) of the water at the inlet and at the outlet.

output of the cooling system. The values are obtained using equations (5) and (6).

- **Temperature variation (ΔT):**

$$\Delta T = Ts - Te \tag{5}$$

Where Ts is the temperature of the water leaving the torch and plasma and Te is the temperature entering the torch.

The value of the temperature change is applied to equation (6):

- **Power dissipated (P_{AB}):**

$$P_{AB} = G_{H2O}.cp_{H2O}.\Delta T \tag{6}$$

Where, cp_{H2O} is the specific heat of water (4.2 kJ/kg.K).

With these values, the thermal efficiency (y) of the plasma torch can be calculated using equation (7).

- **Thermal efficiency (q):**

$$\eta = \frac{P_{total} - P_{AB}}{P_{total}} \tag{7}$$

Once all these values have been obtained, the average enthalpy of the plasma jet can be estimated using equation (8).

- **Average enthalpy of the plasma jet (Δh):**

$$\Delta h = \frac{\eta \, P_{total}}{G_{N_2}} \tag{8}$$

Where G_{N2} is the value of the working gas flow.

1.11. Theoretical calculation of particle velocity and temperature in the plasma jet

In deposition processes such as *cold spray*, the main characteristic is the high particle velocity, but at lower temperatures compared to other thermal spray deposition methods (FAUCHAIS, 2014). The great advantage of this technique is the mechanical adhesion (plastic deformation) obtained due to the high particle speed (SINGH, 2012).

The speed and temperature of the particles in plasma spray processes depend heavily on their interaction with the high temperature of the plasma jet. The characteristics of the particles in flight, before they collide with the substrate, directly influence the quality of the coating. In the conventional plasma spray process, good adhesion of the material to the substrate is achieved when the particle is heated to temperatures intermediate between the melting point and the evaporation point of the material. On the other hand, the particle speed must be high enough to obtain good adhesion, but not so high as to cause it to fragment (FAUCHAIS, 2014; ESSIPTCHOUK, 2017).

In this context, mathematical models can be used to determine the movement and heating of particles in high-velocity, high enthalpy plasma jets (PAWLOWSKI, 2007; FAUCHAIS, 2014). Considering particles of mass mp, their movement is determined by the balance of forces defined in equation (9).

$$F_i = \sum F_j \tag{9}$$

Where the force of inertia $(F_i = m_p \dfrac{dv_p}{dt})$ is given by the product of the mass of the particle (m_p) and the variation of the particle's velocity (dv_p) in time. The inertial force is balanced by the sum of the vectors of the external forces Fj, such as the viscous drag force, the plasma pressure gradient, the drag force due to mass addition (Archimedean), the Basset force (due to the unsteady motion of the particles), the force of gravity, magnetic or electrical forces, turbulence and thermophoresis. In most of the cases studied, the greatest influence is due to the viscous drag force () :

$$F_{d.f} = C_D \frac{\rho_g U^2}{2} A_p \tag{10}$$

Where c_D is the drag coefficient, A_p is the frontal area of the particle, ρ_g is the density of the plasma gas, eis the relative velocity of the particle at relation to the velocity of the u_g plasma jet. Considering the acceleration and deceleration of the particles, this velocity is given by:

$$U^2 = (u_g - v_p)|u_g - v_p| \tag{11}$$

The drag coefficient c_D is a function of the relative velocity of the particle *(U)* *and* can be determined by the Reynolds number, defined by:

$$Re = \frac{(\rho_g d_p)|u_g - v_p|}{\mu_g} \tag{12}$$

Where u_g is the dynamic viscosity of the plasma and d_p is the particle diameter.

Fauchais et. al (FAUCHAIS, 2014) and Pawlowski (PAWLOWSKI, 2007) proposed approximations using $CD = f(Re)$, where for low Reynolds number fluids the expression can be used:

$$C_D(Re) = \frac{24}{Re} + \frac{6}{1 + \sqrt{Re}} + 0{,}4 \tag{13}$$

Thus, the differential equation of motion of a particle in a high-temperature plasma jet is given by:

$$\frac{dv_p}{dt} = \frac{\rho_g}{d_p \rho_p} \left(\frac{18}{Re} + \frac{4{,}5}{1 + \sqrt{Re}} + 0{,}3 \right) (u_g - v_p)|u_g - v_p| \tag{14}$$

Disregarding heat losses due to radiation and phase transformation, the temperature of the particle can be determined using equation (15).

$$m_p c_p \frac{dT_p}{dt} = A_t h_p (T_g - T_p) \tag{15}$$

Where the mass of the particle is given by $m_p = \frac{1}{6}\pi d_p^3 \rho_p$, the area of the particle by $A_t = \pi d_p^2$. The specific heat cp is considered constant and independent of the processing temperature. The heat transfer coefficient hp of a spherical particle is obtained from the Nusselt number:

$$Nu = \frac{h_p d_p}{\lambda_t} = 2 + 0{,}66 Re^{0{,}5} Pr^{0{,}33} \tag{16}$$

The Prandtl number *(Pr)* is obtained from $Pr = \frac{\mu_g c_{pg}}{\lambda_g}$, where its variables are defined according to the properties of the plasma-forming gas for the jet temperature (BOULOS, 1994). The average thermal conductivity of the particle is calculated by:

$$\lambda_t = \frac{1}{T_g - T_p} \int_{T_p}^{T_g} \lambda_t(T) dT \tag{17}$$

According to the assumptions made above, it is possible to define the differential equation for the time variation of the particle's temperature by equation (18).

$$\frac{dT_p}{dt} = \frac{6\lambda_t}{\rho_p c_p d_p^2} (2 + 0.66 Re^{0.5} Pr^{0.33})(T_g - T_p) \tag{18}$$

Finally, equations (14) and (18) form a closed system that makes it possible to theoretically predict the particle's velocity and temperature at any moment in its trajectory in high enthalpy, high velocity plasma jets.

1.12. Description of the experimental procedure

Figure 24 shows a general flowchart of the experimental procedures adopted in this research, considering both the coating processing stage and the material characterization stage.

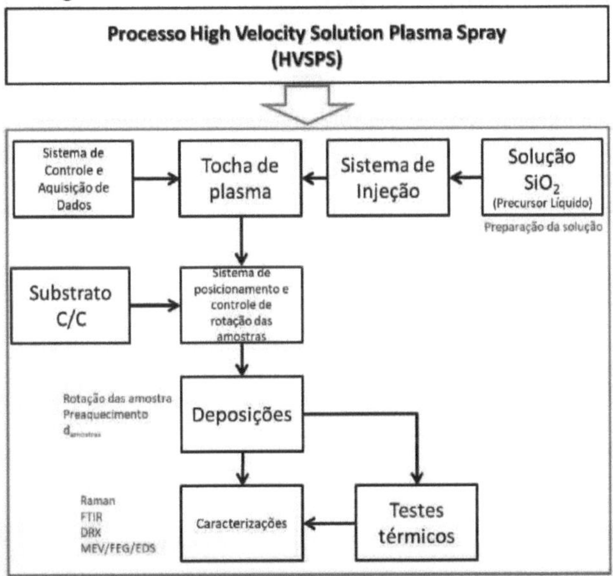

Figure 24. General flowchart of the experimental steps carried out.

In the coating processing stage, preliminary tests were carried out to vary the temperature at which the substrate was preheated, using the plasma torch itself. The results showed that at temperatures below 200°C the coating did not adhere well to the substrate, which led to deposition processes being carried out only above this temperature level. Dynamic depositions were carried out using a positioning system with rotation control. The rotation used was 180 RPM, which is the lowest possible rotation of the sample holder. Subsequently, characterization tests were carried out on the deposited coatings.

4.13. Thermal testing of coatings

The sample coatings were tested in accordance with ASTM standard E285-08 *"The standard test method for the oxyacetylene ablation testing of thermal insulation materials"* ("Standard Test Method for Oxyacetylene Ablation Testing of Thermal Insulation Materials," 2015). Specifically, the purpose of using this test on the coatings was to verify and prove the self-sealing capacity of the silica and possible changes in its atomic structure.

The test described by ASTM E285-08 aims to determine the thermal insulation capacity of ablative materials, using an environment with a constant flow of hot gas supplied by an oxyacetylene torch. This thermal test is commonly used to evaluate the thermal, ablative and microstructural properties of materials or products when exposed to flame under controlled conditions.

In this test, the samples are placed perpendicular to the axis of the flame promoted by the oxyacetylene torch, controlling the pressures of oxygen and acetylene to ensure a stoichiometric mixture between the two gases.

The sample must be positioned at a fixed distance of 20 mm from the flame exit nozzle of the oxyacetylene torch. The temperature of the surface and the back of the sample must be monitored continuously.

This test provides information on the material's erosion rate, which is determined by measuring the thickness of the sample in relation to the time it is exposed to the flame. The effectiveness of the thermal insulation is obtained by measuring the temperature of the back of the sample. The thermal insulation index is obtained by dividing the time the sample reaches temperatures of 80, 180 and 380°C by the original thickness of the sample. Figure 25 shows the layout of the test system components.

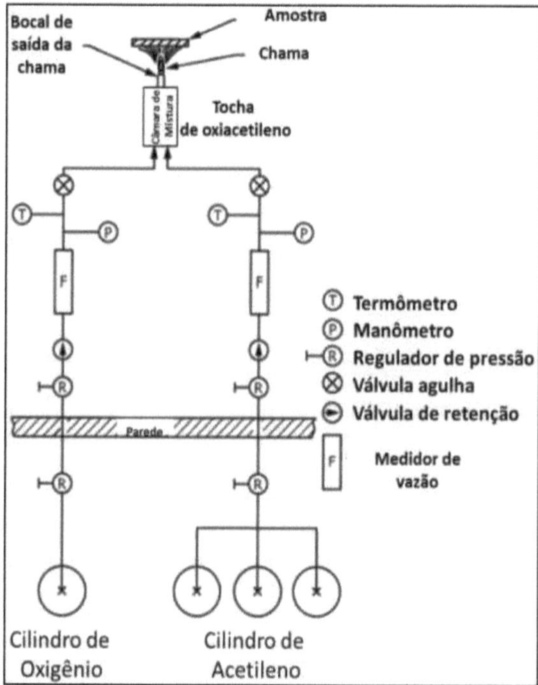

Figure 25. Schematic drawing of the test system components according to ASTM E285- 08 ("Standard Test Method for Oxyacetylene Ablation Testing of Thermal Insulation Materials," 2015).

1.14. Characterization of coatings

1.15. X-ray diffraction

To identify the atomic structure present on the surfaces of the samples, X-ray diffractometry was used, with a copper tube, in a 0/2θ configuration, in Panalytical equipment, model Empyrean, installed in the Materials Characterization Laboratory of the Instituto Tecnológico de Aeronáutica (LabMat/ITA). The analysis parameters were a step of 0.02°, 20 ranging from 15° to 100°, and an interaction time of 10 seconds.

4.16. Scanning Electron Microscopy (SEM/EDS) and Field Emission Scanning Electron Microscopy (SEM-FEG).

To evaluate the microstructural characteristics and the distribution of chemical elements in the synthesized coatings, a Tescan Vega 3 XMU electron microscope from ITA's Materials and Processes Department was used, operating with energy dispersive spectrometry. For higher resolution analyses, another SEM with a field emission *gun (Field Emission Gun-FEG)* of the make/model Tescan/Mira 3, installed at the Associated Laboratory of Sensors and Materials of the National Institute for Space Research (LAS/INPE), was used.

4.17. Raman Scattering Spectroscopy

Raman spectra were collected using Horiba's Evolution equipment, with an excitation wavelength in the visible range (532 nm), installed in the Materials Characterization Laboratory at the Instituto Tecnológico de Aeronáutica (LabMat/ITA). This technique was used to obtain information on the structure and chemical composition of the coatings.

4.18. Infrared spectroscopy (FTIR)

FTIR spectra were obtained from the surfaces and cross-sections of the samples, using a Perkin Elmer Frontier device in ATR mode, with 16 accumulations installed at the Materials Characterization Laboratory of the Instituto Tecnológico de Aeronáutica (LabMat/ITA). This technique was used to identify the chemical composition of the coating.

Chapter 5

5. *High Velocity Solution Plasma Spray* Deposition

High Velocity Solution Plasma Spray (HVSPS) was the name given to the process used in the depositions carried out in this work, which is associated with the characteristics of the plasma torch developed, mainly in relation to the speed of the plasma jet.

5.1. Preliminary studies

The operating parameters used in the depositions were obtained from studies carried out prior to the deposition experiments, within the scope of the work. The operating conditions of the plasma torch were obtained by varying the values of the working gas flow and the operating current of the plasma generator.

The results of the torch characterizations will be presented in Chapter 8 (Results and discussions).

After characterizing the plasma torch, experiments were carried out varying the flow rate of the liquid precursor material, in order to establish the minimum (30 ml/min) and maximum (150 ml/min) flow rates at which the torch operates in a more stable voltage and current regime, as well as sustaining a relatively thick coating synthesis process. For flow rates of less than 30 ml/min, no deposition of the material was observed, suggesting that, under these conditions, there is intense evaporation of the material in the plasma/precursor interaction process. Values higher than 150 ml/min lead to an interruption in the operation of the plasma torch due to an increase in resistivity in the discharge channel. Based on these results, a precursor flow rate of 80 ml/min was chosen for the deposition process, with the possibility of increasing or decreasing this flow rate.

5.2. Deposit tests

The depositions of the coatings were divided into times of 3, 6 and 9 minutes with varying sample distances of 50 and 90 mm from the torch nozzle.

Table 3 shows the identification of the experiment (EXP) and the parameters used in the depositions, which are: deposition time (TEMPO), sample preheating temperature (TEMP. PRE AQUEC), distance between the sample and the torch nozzle (DIST.), solution flow rate (G(solu.)), nitrogen flow rate (G(N_2)), torch voltage (U), torch current (I) and sample rotation (ROT).

Table 3. Parameters used in the depositions

EXP.	TIME [min]	TEMP. PRE HEAT. [°C]	DIST. [mm]	G(Solution) [ml/min]	G(N2) l/min	U (V)	I (A)	ROT. (RPM)
SMP 3 - A	3	200	50	80	250	300	100	180
SMP 3	3	200	90	80	250	300	100	180

SMP 6 - A	6	200	50	80	250	300	100	180
SMP 6	6	200	90	80	250	300	100	180
SMP 9 - A	9	200	50	80	250	300	100	180
SMP 9	9	200	90	80	250	300	100	180

Chapter 6

6.   Results and discussion

Although depositions were made at distances of 50 and 90 mm between the samples and the plasma torch's spray nozzle, an intense process of coating degradation was observed in depositions at a distance of 50 mm, an effect that also recurs in static sample deposition processes. This result is probably associated with the high values of temperature, heat flux and speed of the plasma jet incident on the samples in this position.

In this context, this chapter presents the results regarding the characterization of the plasma torch, the operating parameters used in the deposition process and finally the characterization of the coatings obtained with the substrates kept 90 mm away from the plasma torch nozzle.

6.1.   Plasma torch current-voltage characteristics

The main energy property of the plasma torch is the characteristic that shows the variation of the voltage in the arc as a function of the current with the gas flow as a parameter. Figure 26 shows the characteristic voltage-current curve of the plasma torch for values of 250 and 350 l/min of working gas flow. It can be seen that for a flow rate of 250 l/min the plasma torch operates at a higher current range, between 55-135 A. For the 350 l/min flow, the torch only operates at currents above 70 A. It is worth noting that the maximum current limit obtained is related to the power supply configuration, which is set to operate at a maximum current of 150 A.

There is a tendency for the voltage to vary linearly in relation to the current for the gas flows applied, associated with the characteristics of the power source, which is not purely a current source. The increase in gas flow increases the equivalent electrical resistance of the circuit and consequently the operating voltage (ZHUKOV, 2007).

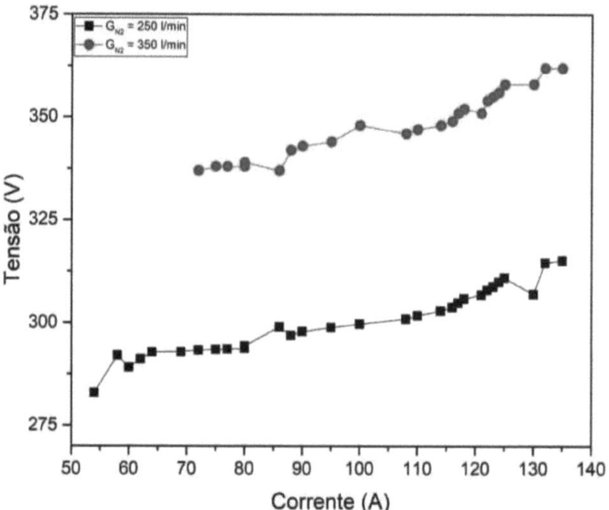

Figure 26. Characteristic curve of the HVSPS process plasma torch. The uncertainty in the measurements is 5%.

The results also show that, unlike commercial torches, the Tandem torch operates at relatively low current values, but at high voltages. This characteristic provides low electrode erosion and minimal plasma contamination, avoiding frequent interruptions to experiments to maintain the torch. In addition to this operational advantage, the low contamination of the plasma by the electrode material also has a direct influence on the physical and chemical properties of the coatings produced.

For higher values of current and lower working gas flow, an increase in the average enthalpy of the plasma jet is observed in accordance with the equations in section 5.10. Figure 27 shows the variation of the average enthalpy of the plasma jet as a function of current, with the working gas flow as a parameter at 250 and 350 l/min. The average enthalpy of the plasma jet varies between 2.5-5 MJ/kg, whereas conventionally, the torches known in the literature used in deposition processes have values between 5-35 MJ/kg (VARDELLE, 2014). Due to the design of the Tandem plasma torch, it is able to process ceramic materials even with lower enthalpy values.

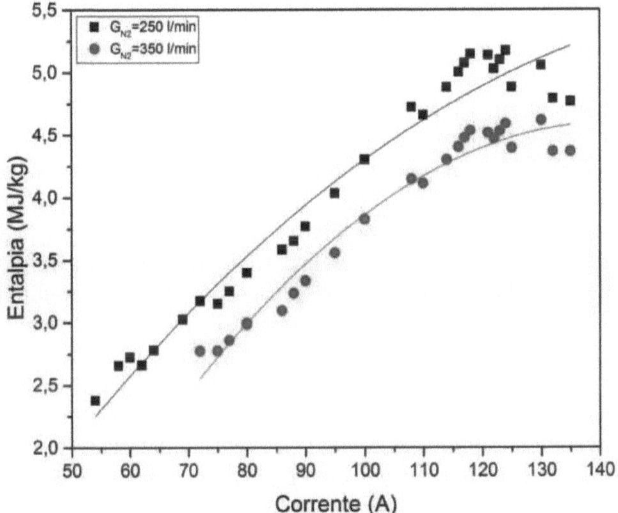

Figure 27. Current enthalpy characteristics for flow rates of 250 and 350 l/min

As shown in Figure 26, the operating voltage varies by around 5% for the gas flow rates used (250 and 350 l/m). Thus, according to Figure 27, we can see an upward behavior of enthalpy when the gas flow and voltage are constant. The enthalpy variation is directly influenced by the variation in the applied current.

Another factor that influences the average enthalpy of the plasma jet is the variation in gas flow. For this comparison, it is necessary to set a current value and evaluate the enthalpy for the different flow rates. Although the flow rates used in the characterizations result in different voltage values, their contribution to the variation in the enthalpy of the jet is smaller when the influence of the gas flow rate is taken into account, as exemplified mathematically by the equation used to calculate the average enthalpy of the plasma jet.

The efficiency of transforming electrical energy into thermal energy in the plasma torch was also evaluated, considering only the heat loss to the cooling system, i.e. obtained from the water flow and temperature measurements at the torch's cooling water inlet and outlet. Thermal efficiency values vary between 75 and 80%, relatively higher when compared to the torches conventionally used in plasma spray processes with values around 30-70% (CALIARI, 2016).

6.2. Numerical study of particle temperature and velocity in flight

Taking into account the parameters of the plasma spray process, the microstructure and quality of the coating are directly associated with two of the main ones, namely temperature and particle velocity. The high rate of plasma-particle heat transfer helps to reduce porosity and the occurrence of unmelted particles in the coating. Although the presence of semi-fused or unfused particles may weaken the coating, a high-speed deposition process causes the particle to deform plastically when in contact with the substrate surface, i.e. the particle adheres mechanically to the coating (KULKARNI, 2003).

Thus, the use of a theoretical approach to the behavior of temperature and particle velocity provided by the HVSPS process helps to understand the mechanisms of formation and nucleation of nanostructures and the properties of the coatings synthesized. This study was carried out using a computational model based on equations 14 and 18, considering solid particles with a defined diameter interacting in the expansion region of the plasma jet. For liquid precursors, the modeling is much more complex because it involves state transformations (liquid to gaseous) and consequent variations in the diameter of the particles that are formed during the processing of the liquid precursor. Nanoscale particles do not have a defined trajectory, which is influenced by thermophoresis forces caused by the temperature and velocity gradient within the plasma jet (FAUCHAIS et al., 2014). For these reasons, we decided to use zirconia partially stabilized with yttria, which has characteristics already determined by its manufacturer in terms of its particle size distribution (d_p = 20 - 60 uni), specific mass (p = 5890 kg/m3) and specific heat (C_p = 466 J/kg.K).

The following assumptions were made for the simulations:

- Two plasma jet velocity values were established, u_g = 400 and 1000 m/s. These speeds were chosen to identify possible variations in the particle's behavior during its heating and acceleration;
- The average temperature of the plasma jet (T_{pl}) was assumed to be 3000 K;
- The particle injection velocity (v_{p0}) varies between 0 -1200 m/s.

Figure 28 shows the simulation results corresponding to the variation of particle velocity in relation to flight distance with particle diameters of 20 and 60 iim and different injection velocity values.

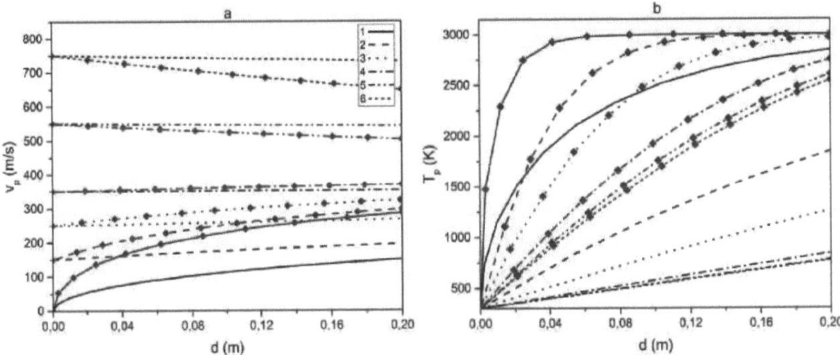

Figure 28. Behavior of (a) particle velocity and (b) particle temperature for plasma jet velocity of 400 m/s as a function of flight distance for particles of 20 µm (curves with symbols) and 60 µm of diameter (curves without symbols). Results obtained for different injection speed values: 0, 150, 250, 350, 550 and 750 m/s, numbered from 1 to 6 in this sequence.

The change in speed (ΔV) is obtained using the equation $\Delta V = \left| v_{p.fin} - v_{p0} \right|$, where vpfin is the particle's final velocity and vp0 is the particle's injection velocity. It was observed that the greatest variation in velocity is obtained for the injection velocity $v_{p0} = 0$. During the flight distance of 0.2 m, the particle's final velocity increases to 150 m/s, corresponding to 38% of the plasma jet's velocity.

As the particle injection velocity increases, the velocity variation (Av) decreases until it reaches zero (for $v_{p0} = u_g$) and in this condition, the drag force is also zero. For v_{p0} greater than the plasma jet velocity, the final velocity is not significantly affected. For example, for $v_{p0} = 750$ m/s, the decrease in the initial velocity of the particle with a diameter of 60 ^m is relatively small and reaches a value of $Av = 20$ m/s, or 5% of the plasma jet velocity.

Figure 28(a) shows that the variation in velocity (Av) increases for particles with a smaller diameter. For a diameter of 20 ^m and vp0 = 0, the velocity increases up to $v_{pfin} = 285$ m/s (which corresponds to 71% of the plasma jet velocity). However, when the particle injection velocity is $v_{p0} = 750$ m/s, the final velocity of the particle decreases by 15%.

The behavior of the particle temperature in relation to the distance from the injection nozzle is shown in Figure 28(b). For lower values of particle injection velocity, an asymptotic increase in particle temperature (T_p) was observed. Small particles can be heated to temperatures close to the plasma temperature over relatively short flight distances. As v_{p0} increases, the heat transfer coefficient and flight time decrease and the particle needs longer flight distances to reach temperatures close to those of the plasma jet. This effect is intensified as the particle diameter increases.

From differential equations 14 and 15, it can be concluded that both the acceleration of the particle and the temperature are associated with the relative velocities of the plasma jet and the particle. The acceleration of the particle is directly proportional to Av, if, $(\frac{d_{vp}}{dt} \sim \frac{\Delta v}{d_p^2})$, inversely proportional to the diameter of the particle. While

temperature of the particle is less influenced, i.e. $\frac{d_{Tp}}{dt} \sim \frac{|\Delta v|^{0.5}}{d_p^{1.5}}$. The acceleration or The deceleration of the particles depends on the sign of Av and increases with Av. Therefore, particles with zero injection velocity have greater acceleration than those with values greater than zero. However, in the case of $vp0>ug$, the particle's inertia is sufficiently high, while the time of its interaction with the plasma jet is short, resulting in a gentle reduction in the particle's initial velocity.

The competition between the rate of change of speed and the rate of heating of the particles results in a rapid decrease in the final temperature of the particles at a fixed flight distance. The rate of acceleration and heating of the particles are inversely proportional to the diameter of the particles at powers 2 and 1.5, respectively. Thus, under the conditions studied, particle diameter has a greater influence on particle velocity than on temperature.

Figure 29 shows the results of the final velocity and temperature of the particle in relation to the injection velocity and the particle diameter for plasma jet velocities of 400 and 1000 m/s.

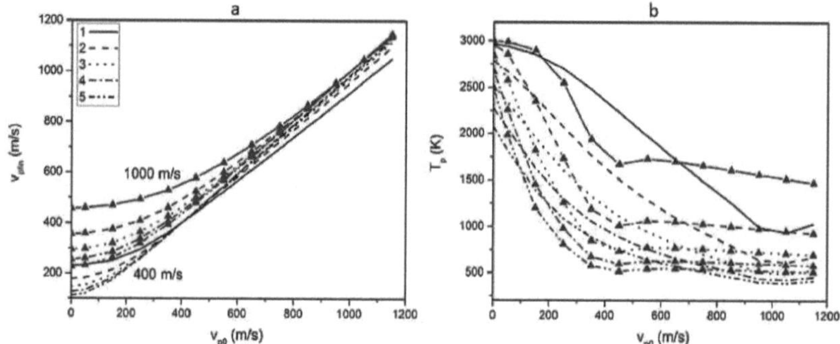

Figure 29. Velocity (a) and temperature (b) of the particles at a distance of 0.1 m from the injection point as a function of the particle injection velocity for a plasma jet velocity of 400 m/s (curves without symbols) and 1000 m/s (curves with symbols) and different particle diameters: 20, 30, 40, 50 and 60 ^m numbered from 1 to 5 in this sequence.

In all cases, regardless of the velocity of the plasma jet, the increase in injection velocity results in an increase in the final velocity of the particle which follows a linear behavior, as can be seen in equation 14. The greatest dispersion of the particle's final velocity values in relation to its diameter occurs at low injection velocities, with the greatest variation observed for $v_{p0} = 0$. As v_{p0} increases, the dispersion of final velocity values decreases until all particles with different diameters reach the same velocity. This occurs when $v_{p0} = ug$ and the drag force disappears. In the case of $vp0>ug$, a new dispersion of velocity values is observed, but with less intensity.

This result shows the advantage of using high injection speeds to obtain a flow of particles with different diameters and a more uniform distribution of speeds   .

Figure 29 (b) shows the results of the variation in particle temperature in relation to the distance (0.1m) traveled in the plasma jet, for different particle diameters.

A sharp drop in particle temperature is clearly observed when $v_{p0} < u_g$. The lowest particle temperature is reached for $v_{p0} = u_g$, where heat transfer occurs only by conduction (which is indirectly taken into account in this model), with less intensity compared to forced convection. At the point where $v_{p0} = u_g$, it can be seen that the injection speed has little influence on the particle's temperature.

It is also interesting to note that, after a certain value of v_{p0}, the temperature of the decelerated particle (for $v_{p0} > u_g$ in a plasma jet with $u_g = 400$ m/s) exceeds the temperature of the accelerated particle (for $v_{p0} < u_g$ in a plasma jet with $u_g = 1000$ m/s). This result shows that the particle heating process, at high injection speeds and at certain flight distances, is more effective in the plasma jet with lower speeds. In the case studied, the critical value of the injection velocity is between 650 and 700 m/s.

Figure 30 shows the influence of the particle's diameter on its temperature and speed.

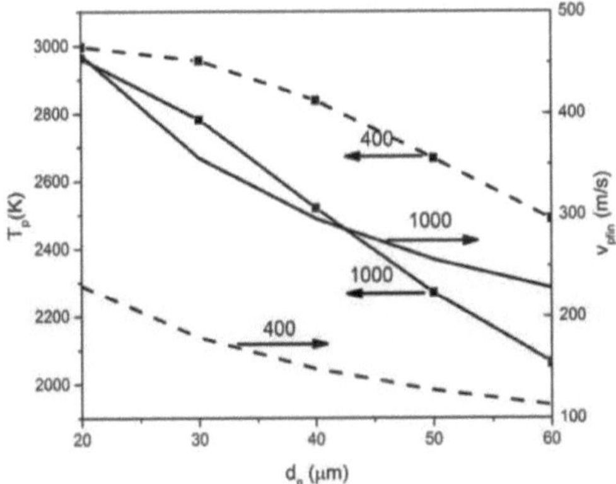

Figure 30. Particle velocity and temperature at a distance of 0.1 m from the injection point as a function of particle diameter and particle injection velocity for different
plasma
jet velocities.

To obtain the results shown in Figure 30, the particle's injection velocity was assumed to be $v_{p0} = 0$. It can be seen that increasing the particle diameter decreases both its speed (due to the increased mechanical inertia) and its temperature (due to the increased thermal storage capacity). Furthermore, for the case of a high-speed plasma jet, all the particles undergo intense acceleration, while the temperature is lower compared to a low-speed jet. The final particle velocity v_{pfin} is reached at a flight distance of 0.1 m and decreases by 50% when the diameter is changed from 20 to 60 pm, regardless of the plasma jet velocity. However, the final temperature of the particles depends strongly on the speed of the jet. The increase from 400 to 1000m/s results in an increase in the temperature difference between particles of diameters d_1 and d_2 $[\Delta T = T(d_1) - T(d_2)]$. For the diameters of particles $d_1 = 20\ \mu m$ *and* $d_2 = 60$ pm, the temperature difference varies from 17% (for $u_g = 400$ m/s) to 30% (for $u_g = 1000$ m/s).

For comparison, instead of using the absolute values of particle velocity v_p and temperature T_p, the heating and acceleration processes can be analyzed in terms of kinetic energy $(E_c = \frac{m_p v_p^2}{2})$ and thermal energy ($E_T = m\ c_{pp}\ (T_p -$) of the particle. In addition, in order to generalize the results, the specific energy values per unit mass are applied, i.e. the specific kinetic energy $(e_c = \frac{v_p^2}{2})$ and specific thermal energy ($T_T = c_p\ (T_p - T_{p0})$), both in J/kg.

Figure 31 shows the effect of particle injection velocity on the specific energies (kinetic, thermal and total, where) for velocities
of the plasma jet at 400 m/s (Figure 31a) and 1000 m/s (Figure 31b). Again, two particle diameter values were set (dp = 20 and 60 pm). The results show that kinetic energy increases with particle injection speed and decreases with thermal energy. Under these conditions, the lowest total energy value was obtained.

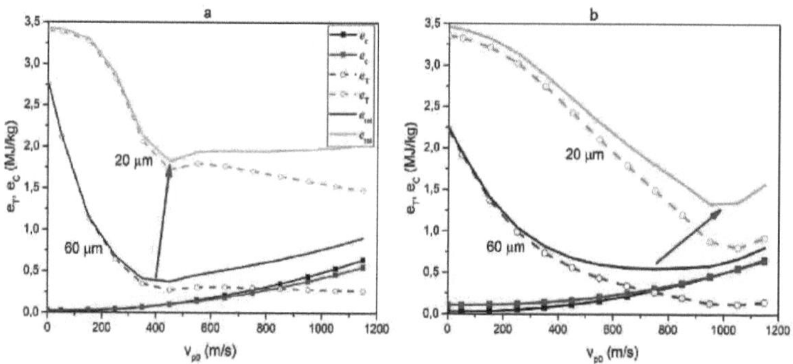

Figure 31. Particle specific energy as a function of particle injection speed at a distance of 0.1 m from the injection point for different particle diameters at ug = 400 m/s (a) and 1000 m/s (b).

For a plasma jet speed of $u_g = 400$ m/s, the lowest total energy is observed when $v_{p0} = u_g$, varying according to the particle diameter.

Figure 31a shows that the specific kinetic energy for two values of particle diameter are very close, and the same dependence is observed for the high-speed plasma jet. As the particle diameter increases, the total energy changes little when lower values of v_{p0} are considered.

To estimate the effect of plasma velocity and particle diameter on the total specific energy, two parameters are introduced. The first is defined in the equation:

$$\alpha_{\frac{d_{p1}}{d_{p2}}} = \frac{(e_{tot}|d_{p1})}{(e_{tot}|d_{p2})} \tag{19}$$

Equation 19 represents the total specific energy of the particle with diameter d_{p1} in relation to the energy of the particle with diameter d_{p2} in a plasma jet with constant velocity u_g.

Another parameter used is defined in equation 20.

$$\beta_{\frac{u_{g1}}{u_{g2}}} = \frac{(e_{tot}|u_{g1})}{(e_{tot}|u_{g2})} \tag{20}$$

This equation relates the total specific energy of the particle of diameter dp obtained in a plasma flow with a velocity u_{g1} to the total energy obtained by the same particle in a plasma flow with a velocity u_{g2}.

Figure 32 shows the graph of the influence of the particle's initial velocity on the parameters α e β.

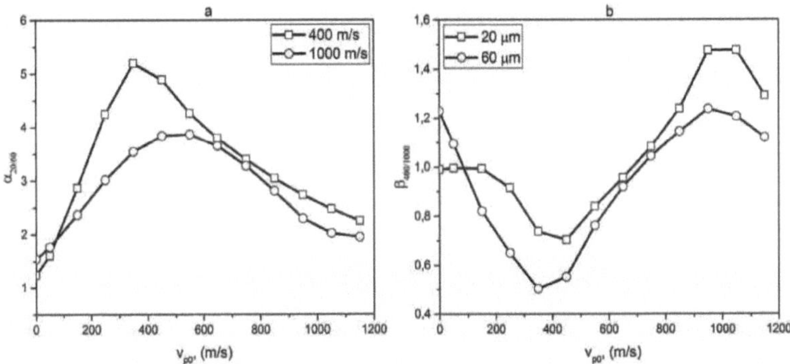

Figure 32. Dimensionless parameters a (a) and P (b) as a function of particle injection speed for different plasma jet speeds (a) and different particle diameters (b).

For the calculations we used the particle diameters $d_{p1} = 20\,\mu m$ *and* $d_{p2} = 60$ ^m and the plasma jet speeds $u_{g1} = 400$ m/s and $u_{g2} = 1000$ m/s, setting the particle flight distance at 0.1m. From Figure 32(a) it can be seen that the energy of the particle with a diameter of 20 ^m exceeds the energy of the particle with a diameter of 60 μm $(\alpha>1)$ with the maximum value of α equal to 5 for $u_g = 400$ m/s and almost 4 for $u_g = 1000$ m/s. Both maximum values are located between the injection speeds of 350 and 550m/s. The increase in u_g decreases the a-parameter and changes its maximum value for the higher values of v_{p0}.

Figure 32(b) shows the effect of v_{p0} on the parameter β. It is clear that, for the injection speed range of 100-700 m/s, the plasma jet with $u_g = 1000$ m/s contributes much more to the total energy of the particles $(\beta <1)$. Outside this range, the lower speed plasma jet is still more effective in increasing the total energy of the particles. This result is explained by the greater absolute difference between the velocities of the plasma and the particles, $|u_g - v_p|$.

The change in the total specific energy of the particles in relation to the particle injection speed is shown in Figure 33.

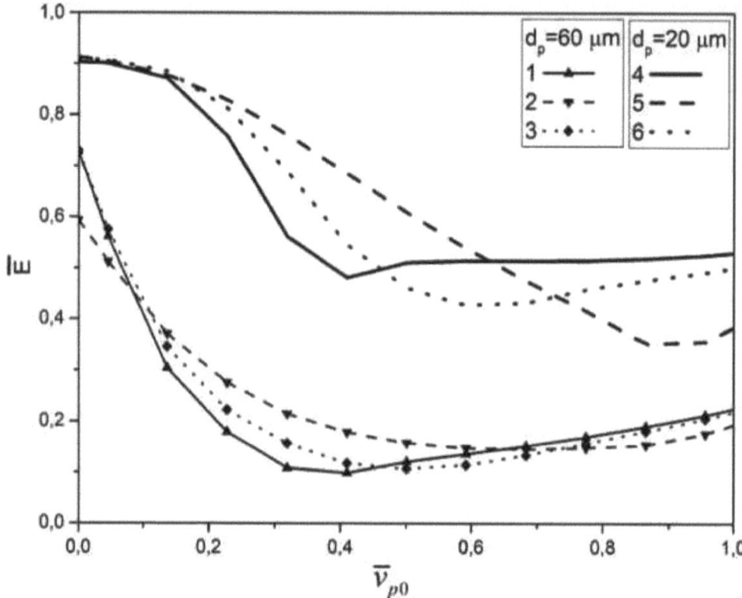

Figure 33. Normalized particle energy as a function of particle injection velocity at a distance of 0.1 m for different plasma jet velocities: 1- 4= 400 m/s; 2- 5=1000 m/s and 3- 6 corresponding to the linear velocity increment from 400 to 1098 m/s.

The enthalpy of the plasma jet, h = 3.79 MJ/kg for T = 3000 K (BOULOS, 1994), was adopted as the basis for calculating the normalized energy A $\overline{E} = \frac{e_{tot}}{h}$. speed of sound at air plasma temperature ($_{aar}$= 1098 m/s) was used as the basis for the normalized velocity calculations $(\overline{v_{p0}} = \frac{v_{p0}}{v_{ar}})$.

The particle diameters, 20 and 60 µm, form two sets of data as shown in Figure 33. Each data set shows the normalized energy of the particles, modelled for constant plasma jet velocities of 400 and 1000 m/s, as well as for the plasma velocity increasing linearly from 400 m/s (at the point of injection) to the speed of sound at a distance of 0.1 m. It can be seen from the results that higher values of total particle energy are achieved only if the initial particle velocity is equal to zero. The particle diameter has the greatest influence on the energy compared to the particle injection velocity.

Finally, with regard to the model considered, it can be concluded that the increase in the initial velocity of the particle for a plasma jet with constant velocity and temperature significantly reduces the temperature of the particle which, in turn, determines its total energy. When the particle injection velocity is greater than the plasma jet velocity, the contribution of kinetic energy tends to match thermal energy (for particles with small diameters (20 iim) or even exceed it (in the case of particles with larger diameters (60 uni)). For the model used, the particle reaches the highest total energy (kinetic and thermal) only when the injection velocity is equal to zero.

1.3. Monitoring process parameters

Monitoring the parameters during the deposition process is extremely important for the reproducibility of the experiments and also for analyzing the properties of the coatings. Figure 34 shows the variation of the main process parameters (substrate temperature, pressure, voltage and torch current) in relation to the deposition time of the coatings on the SMP 3, SMP 6 and SMP 9 samples, delimiting the preheating time and the deposition process by dashed lines.

The surface temperatures of the samples remained the same regardless of the process. Preheating takes an average of 75 seconds to reach a temperature of 200°C. For longer periods of time, with the start of the injection of precursor material, it can be seen in all processes that the temperature undergoes a gentle increase until it reaches 350°C at around 250s.

The pressure inside the torch behaved differently in the SMP 3 experiment, with a sudden increase in pressure due to the total clogging of the torch nozzle caused by the accumulation of material in this region. In the other deposition processes, this problem was not observed, probably due to the operating parameters of the plasma torch, which act to degrade the material accumulated in the nozzle. Naturally, the effect of material deposition inside and on the nozzle of the plasma torch is always present in all processes, leading to a gradual increase in pressure inside the torch in relation to time.

With regard to voltage, the SMP 3 deposition process showed oscillations of around 10%, most likely due to the increase in internal pressure and the consequent variation in arc dimensions. The other processes maintained voltage oscillations of around 6%, with higher voltage peaks being observed from 5 minutes into the process. This effect is related to the accumulation of material in the nozzle which momentarily increases the pressure inside the torch, affecting the dimensions of the arc.

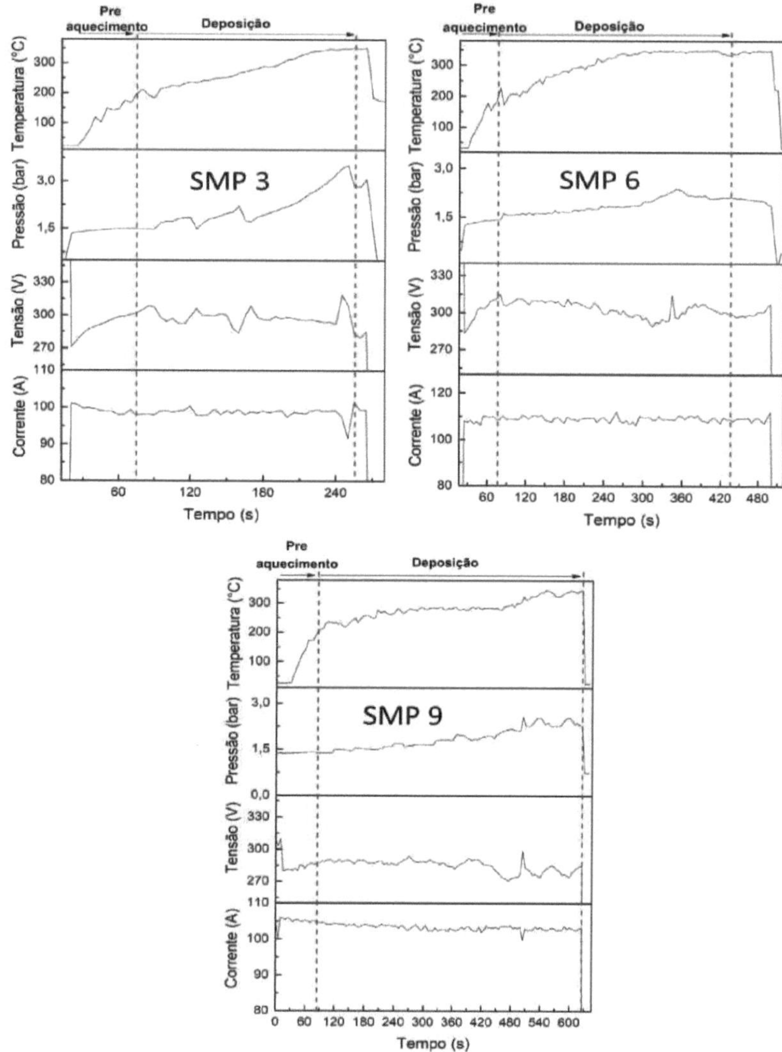

Figure 34. Processing parameters (current, voltage, pressure and substrate surface temperature) for samples SMP 3, SMP 6 and SMP 9 as a function of sample exposure time.

Finally, it was observed that the operating current of the plasma torch remained practically stable during all the deposition processes. The plasma torch operates in relatively stable regimes at voltage and current levels even when the liquid precursor is injected into it, this process being the most critical moment of operation. However, due to the various experiments carried out, it has been possible to establish a procedure to minimize these arc disturbances:

- Adjust the pressures in the tank and the atomizing gas line so that the pressure in the tank is slightly higher than the pressure in the atomizing gas line. In the experiments, pressures of 4.5 and 5 bar were used in the atomizing gas line and the reservoir, respectively.

- Ignite the torch initially only with gas flow from the atomizer (60 l/min) without injecting the liquid precursor.
- Slowly open the valve that releases the injection of the liquid precursor after the preheating process.
- Slowly increase the flow rate of the liquid precursor to the desired flow rate of 80 ml/min.

1.4. Characterization of the coatings

The following sections present the results of the characterization of the coatings obtained using the HVSPS technique, at an average deposition rate of 6.4 Lim/min correlated with the deposition times used of 3, 6 and 9 minutes.

1.4.1. Chemical and structural analysis

1.4.1.1. Raman spectroscopy

Figure 35 shows the Raman spectrum of the C/C substrate used without coating. It is worth noting that no bands were identified in this spectrum between 200 and 1200 cm^{-1} , with only the presence of two peaks located at 1340 and 1580 cm^{-1} , which respectively correspond to the D and G bands characteristic of carbonaceous materials.

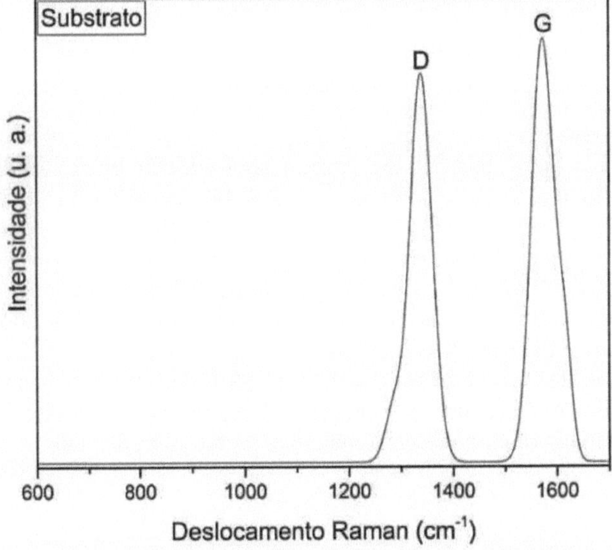

Figure 35 Raman spectrum of the C/C substrate without coating

Figure 36 shows the results of the Raman spectra (black line) and also the deconvoluted curves (colored lines) of the surface of the SMP3, SMP6 and SMP 9 samples after the deposition process, respectively at 3, 6 and 9 minutes.

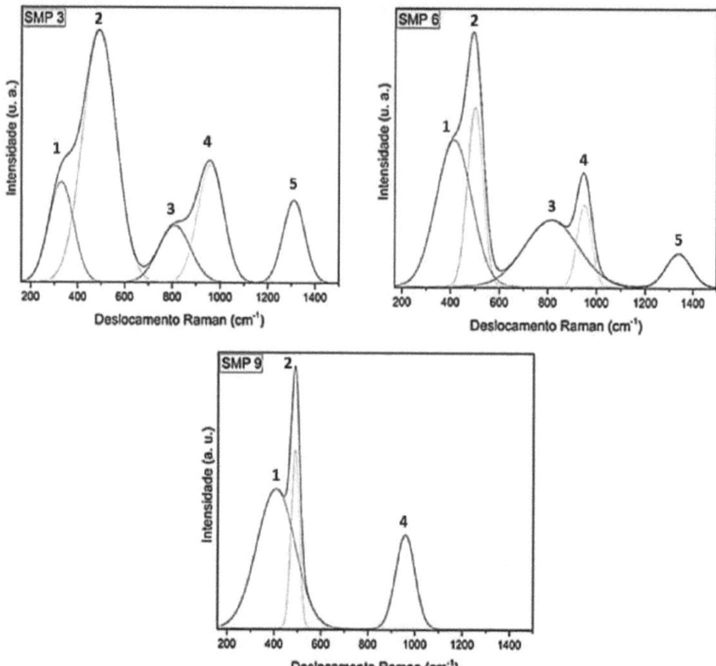

Figure 36. Raman spectra of the SMP3, SMP 6 and SMP 9 samples obtained after the deposition process.

The results show that all the spectra showed two broad bands (1 and 2) corresponding to the Si-O-Si bond at 400 and 490 cm^{-1} (BERTOLUZZA, 1982; BANSAL, 2013). The broadening of the band indicates the structural disorganization of the coating, a typical characteristic of amorphous materials. The samples also show a band (4) located at 960 cm^{-1} , which is associated with the stretching of the Si-OH bond, which was expected as it is a typical bond in the structure of the solution used in the depositions (DECOTTIGNIES, 1978).

Looking at the spectra of the SMP 3 and SMP 6 samples, we can also see the presence, but with less intensity, of band (5) at 1340 cm^{-1} , which is associated with the D band characteristic of carbonaceous materials such as the substrate.

It is also possible to observe a band (3) located at 800 cm^{-1} which is associated with the Si-C bond. The SiC formation reactions are probably due to the reaction:

$$SiO_{(s)} + 2C_{(s)} => SiC_{(s)} + CO_{(g)}.$$

Where the solid SiO formed in the plasma jet reacts with the carbon in the substrate to form solid SiC and CO gas.

This result is not commonly observed in research related to these materials, since the entire process is carried out at atmospheric pressure. Figure 37 shows schematically the probable process of SiC formation during the depositions, starting from the liquid precursor processed by the plasma torch.

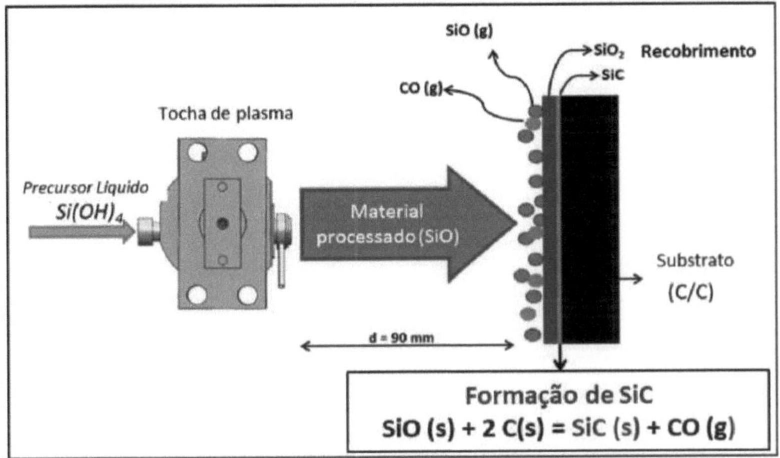

Figure 37. Formation reactions of SiC in the coating during the deposition process.

When comparing the results of the characterization of the synthesized coatings, specifically the SMP 9 sample, with the pure substrate, it can be seen that the peaks referring to SiC and the characteristic carbon bands disappear. This is due to the thickness of the coating and the depth at which the equipment analyzes the material. Although the coating is uniform throughout the substrate, the Raman analysis is very punctual. All the samples were scanned for five times at different points to check for changes between the regions investigated, but all the measurements remained very close, even in samples from different batches. From a process point of view, these results contribute to certifying the reproducibility of the properties of the synthesized coatings.

1.4.1.2.   Infrared spectroscopy (FTIR)

Figure 38 shows the FTIR spectra (black line) and also the deconvoluted curves (colored lines) of the SMP3, SMP 6 and SMP 9 samples obtained with the spectrometer operating in absorbance mode.

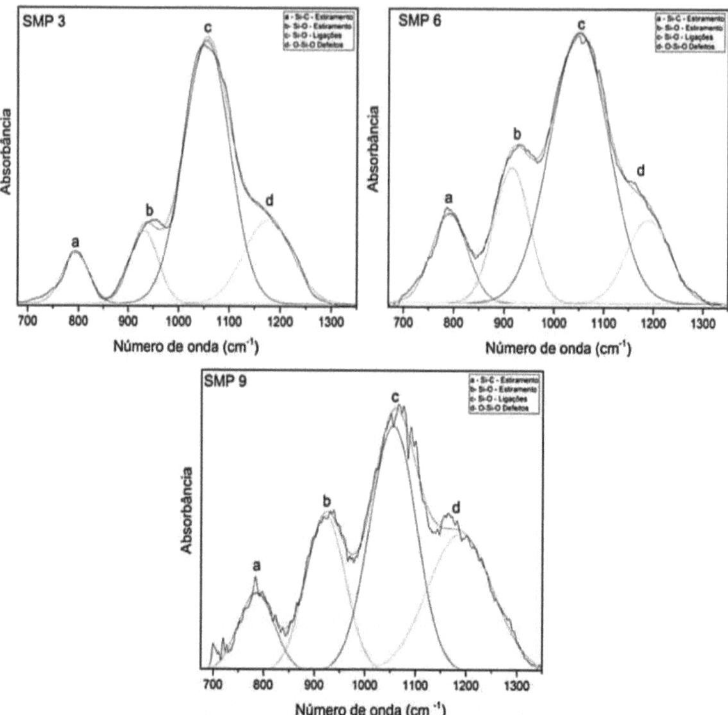

Figure 38. Absorbance spectra of samples SMP3, SMP6 and SMP9.

Four vibrational modes are observed in all the spectra. The main band located at approximately 1050 cm^{-1} is related to the stretching of the Si-O bond (c), at 950 cm^{-1} to the Si-O bond (b) and at 1200 cm^{-1} to the defects present in the Si-O bond (d). These defects are formed mainly by the presence of oxygen at the SiO_2 /SiC interface (GUO., 2001; BARBOUCHE, 2017). The band located at 800 cm^{-1} is related to the Si-C bond (GUNDIAH, 2002b) confirming the formation of SiC during the deposition process. The formation of SiC is related to the partial reduction of silica, probably associated with the relative intensities of the absorbance peaks obtained, where the main peak of the Si-O bond (c) decreases proportionally to the increase in Si-C (a) with increasing exposure time. As the relative intensities or areas of the relative deconvoluted curves of the peaks in the FTIR spectrum are proportional to the concentration of a given material (HUNT, 2016), the increase in SiC formation can be determined by the ratio of the areas of the deconvoluted curves in relation to the main Si-O (c) absorption peak in relation to Si-C (a), using equation 21.

$$P_A = \frac{A_{SiC}}{A_{SiC} + A_{SiO}}$$
(21)

Where, A_{SiC} is the total area of the deconvoluted curve referring to the Si-C absorption peak and A_{SiO} is the total area of the deconvoluted curve referring to the Si-O absorption peak.

For the SMP 3, SMP 6 and SMP 9 samples, the area ratios were 0.1, 0.16 and 0.2, respectively. These results indicate an increase in the formation of Si-C bonds with increasing deposition time and, consequently, with exposure time to the plasma jet. Similarly, the area related to defects at the SiO_2/SiC interface (d) increases, suggesting that Si-O bonds predominate with exposure time and the consequent increase in coating thickness.

Compositional gradient of coatings

The formation of SiC due to precursor/plasma/substrate interaction was demonstrated from the results of the Raman and FTIR spectra on the surface of the coating. To verify the synthesis of a coating with a compositional gradient between SiO2 and SiC along the coating cross-section, the samples were cold-embossed and sanded. Again, the Raman spectroscopy technique was used, taking measurements in three regions from the cross-section of the coated samples. As shown in Figure 39, measurements were taken along the cross-section of the coating, with a spacing of 30 ^m between each point or position, and a series of 15 measurements was taken. The points are located at the interface (1) between the substrate and the coating, the intermediate region (2) and near the surface of the coating (3) exposed to the environment.

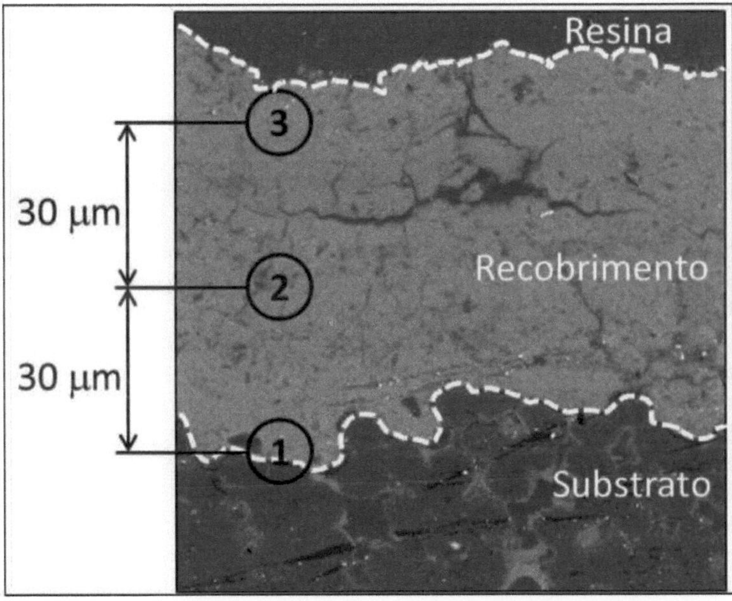

Figure 39. Composition analysis regions where: (1) interface between substrate and coating, (2) intermediate region and (3) coating surface exposed to the environment.

Figure 40 shows the Raman spectra of the samples in the different regions analyzed.

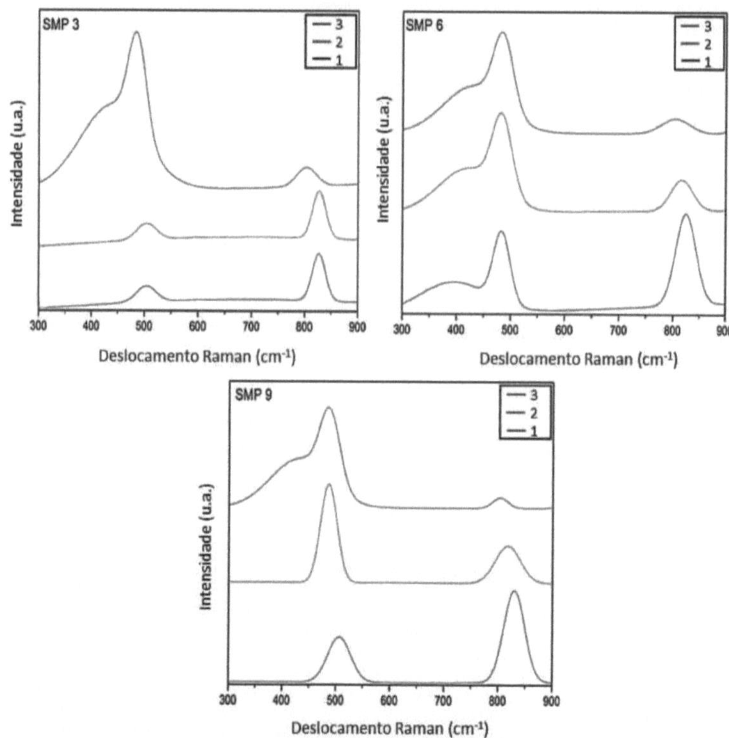

Figure 40. Raman spectra of samples SMP 3, SMP 6 and SMP 9 in regions 1 (-), 2 (-) and 3 (-).

All the samples, regardless of the deposition time, show peaks related to the Si-O-Si bonds (at approximately 500 cm^{-1}), with this peak having a lower intensity at position 1 (substrate/coating interface) when compared to positions 2 (intermediate region) and 3 (near the surface). In an opposite behavior, the peaks related to the Si-C bonds (at approximately 820 cm^{-1}) decrease in intensity as you move away from point 1 (interface) towards point 3 (surface) of the coating. With these results, it can be concluded that the coating obtained has a compositional gradation between SiO_2/SiC, due to the interaction reactions between the C/C substrate and the plasma-sprayed material during the deposition process. When the peak areas of the deconvoluted curves are compared, a higher concentration of Si-C bonds is seen at the substrate/coating interface. The concentration of Si-C bonds decreases as one approaches the outer layer (surface) of the coating, while the concentration of Si-O bonds increases, which is to be expected as silica is the predominant material near the surface of the coating exposed to the environment.

The formation of SiC in the coating helps to increase the chemical and thermal compatibility between the substrate and the coating, improving adhesion (HALD,

2003). Some studies report the need for a step prior to depositing the coating on carbon-based composite substrates. More precisely, SiC depositions are carried out to increase the compatibility between the substrate and the coating material, called bond coatings. Such a procedure was not necessary in the depositions carried out in this work using the HVSPS technique, since the formation of SiC occurs spontaneously in the process of interaction between the plasma-sprayed liquid precursor and the substrate.

Qi et al (QI., 2014) present reactions similar to those that occurred in the samples obtained in the study reported in this thesis, but in this work the SiC is obtained after 6 hours in a furnace with a controlled atmosphere of argon at 1420°C, using the carbothermal chemical vapor deposition technique. The aim was to deposit a SiC coating on carbon nanotubes. Gundiah et al (GUNDIAH, 2002a) described the SiC formation reaction in a controlled atmosphere of NH3 or H_2, using silica gel mixed with activated carbon, with the reactions taking place at a temperature of 1360°C for a period of 4-7 hours. Finally, another study related to the formation of SiC is described by Allaire et al (ALLAIRE et al., 1991), aimed at obtaining SiC powder using a direct current torch, responsible for initiating the reactions between methane and silicon tetrachloride gases. These works show the complexity of precursor manufacturing systems for obtaining SiC, which in most cases require systems with atmosphere control and specific gases, unlike the HVSPS process presented here, which synthesizes SiC at atmospheric pressure in times of less than 9 minutes.

1.5. Atomic structure analysis

1.5.1. X-ray diffraction

Figure 41 shows the X-ray diffraction results of the substrate and the coatings obtained by the HVSPS process.

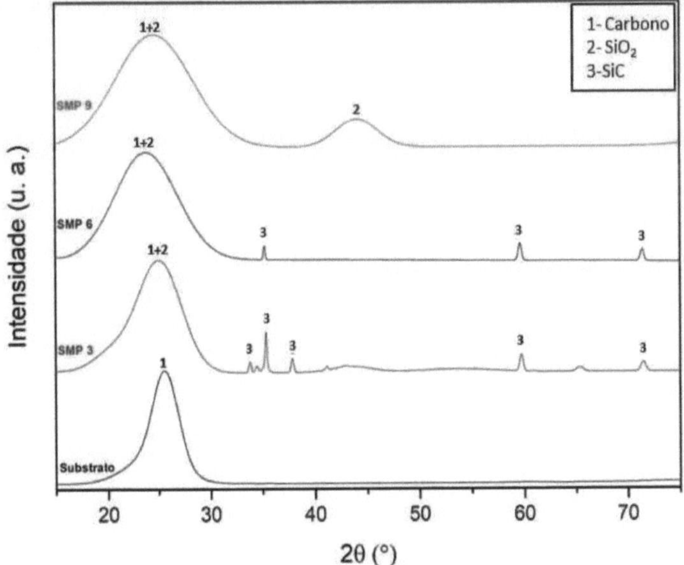

Figure 41. X-ray diffraction spectra of the substrate and coatings of the SMP3, SMP 6 and SMP 9 samples.

Comparing the spectra, it can be seen that the substrate (black line) only contains a diffraction peak at 25°, related to carbon. In the coated samples, new diffraction peaks are observed, especially in the SMP 3 sample (red line), located at 33°, 35°, 37°, 60° and 71°. These peaks are related to the formation of hexagonal SiC (ABOURIDA, 2014; YANG, 2016). The halo located at the 26° diffraction angle indicates the strong presence of SiO_2, which is the main component of the coating. Its widening (halo) indicates a lack of crystalline periodicity, i.e. a strong presence of amorphous material (QUINN, 2005; FAHRNER, 2013). SiO_2, when cooled rapidly, favours the formation of amorphous material, since obtaining crystalline material requires cooling slowly enough for it to organize (XIE, 2004; PAWLOWSKI, 2009; FAUCHAIS, 2014). The formation of amorphous SiO_2 is very advantageous when considering the type of application suggested in this study, which is related to use in environmental protection systems. Due to the characteristic disorganization of amorphous materials, in which the component atoms do not have a periodicity, there is consequently a much lower heat transfer rate when compared to crystalline materials, reducing thermal diffusivity through the coating (MOON, 2002).

In the spectrum obtained after analyzing the SMP 6 sample (blue line), we can see the presence of the halo at 26° related to SiO_2, and again peaks related to hexagonal SiC, but the peaks with lower intensity have disappeared, an effect probably related to the greater thickness of the coating. Finally, the spectrum of sample SMP 9 (pink line) is evaluated. This spectrum does not show the presence of the peaks related to SiC, but only the halo (26°) related to SiO_2. However, there is a second halo located at approximately 44° which also indicates the presence of SiO_2 . The predominance of SiO_2 in this spectrum is directly related to the greater thickness of the coating obtained in this set of samples, which were deposited for a longer time (9 minutes) and the absence of SiC, obviously due to the distance between the surface interaction of the plasma jet and the substrate.

1.6. Microstructural analysis

1.6.1. Scanning Electron Microscopy/EDS/FEG

Figure 42 shows the top images of the coatings obtained in the experiments carried out.

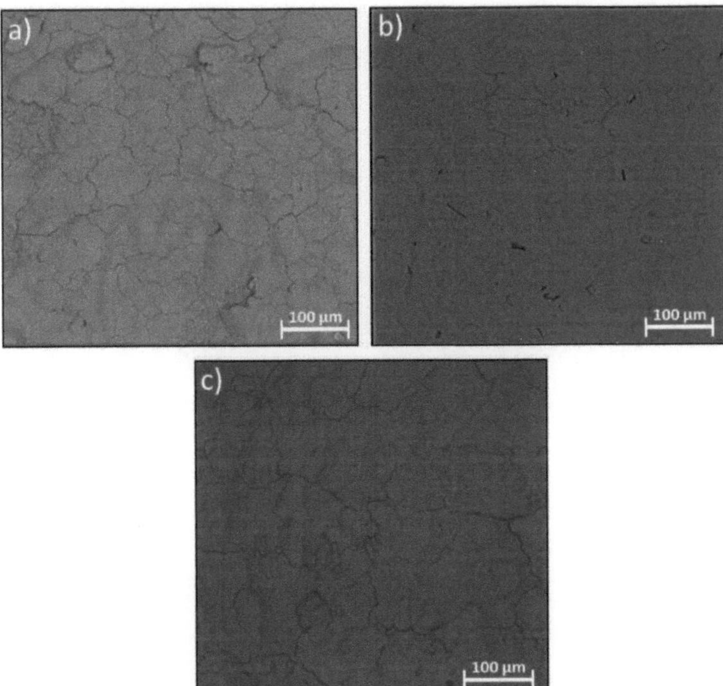

Figure 42. Top views of the coatings on samples (a) SMP 3, (b) SMP 6 and (c) SMP 9.

It can be seen in all the samples that the coatings have cracks distributed throughout their length. These cracks are due to the difference between the coefficient of thermal expansion of the C/C substrate (6.0 x 10-6/°C) and the deposited material (SiO2 (1.0 x 10-6/°C)). However, when exposed to a high-temperature oxidizing environment (ablative), the silica remelts, sealing the cracks and inhibiting oxidative degradation of the composite. This ability to self-fill cracks is referred to in the literature as self-healing and is an intrinsic characteristic of the silica used to benefit the thermal protection system (ZWAAG, VAN DER, 2010). In order to better evaluate the effects of self-healing and its ability to protect in oxidizing environments at high temperatures, it is necessary to thermally test the samples. The results obtained through these tests will be reported in section 8.7 of this thesis.

Before carrying out the thermal tests in an ablative environment, the samples were cold-embedded so as not to compromise their integrity or mask any results, since conventional embeds use heat presses which can in some way interfere with the physical and chemical properties of the coating. Therefore, the samples were prepared following the metallography methodology with the exception of the polishing stage.

The results of the prepared samples are shown in Figure 43. It can be seen that there are not only cracks on the surface of the sample, as shown above, but also cracks in the horizontal and vertical directions inside the coating.

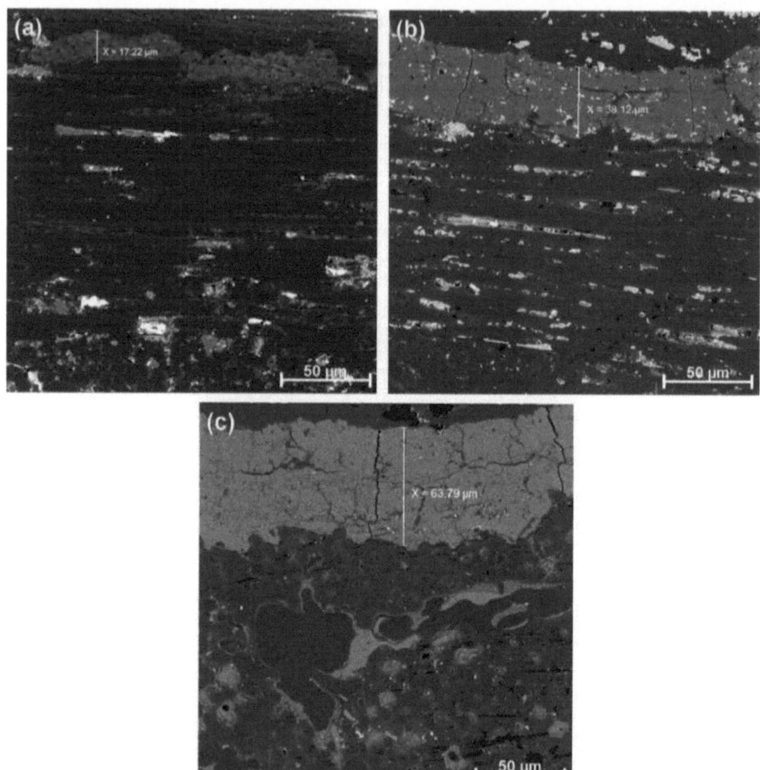

Figure 43. Cross-section of the coatings of samples SMP 3, SMP 6 and SMP 9

Using the images obtained, it is also possible to measure the thickness of the coatings, although they are irregular, so that the deposition rate can be calculated by relating the thickness of the coating to the deposition time. On average, the thicknesses of the coatings are 17.2 µm, 38.1 pm and 63.8 µm corresponding to deposition times of 3, 6 and 9 minutes, respectively. Under these conditions, deposition rates are estimated at 5.7 pm/min for sample SMP 3, 6.4 pm/min for sample SMP 6 and 7.1 pm/min for sample SMP 9. This results in an average deposition rate of around 6.4 pm/min.

As it is a very porous substrate, the deposited material not only covers but also penetrates the composite, contributing to the adhesion of the coating. As shown in Figure 44, another cross-sectional image of the SMP 9 sample was taken in order to better observe how the sprayed material penetrates the composite, filling the voids. An EDS analysis was also carried out in "line scan" mode to once again check the coating composition and assess which region has the highest elemental count. The figure also illustrates the regions with predominant concentrations of SiO_2, SiC and Carbon.

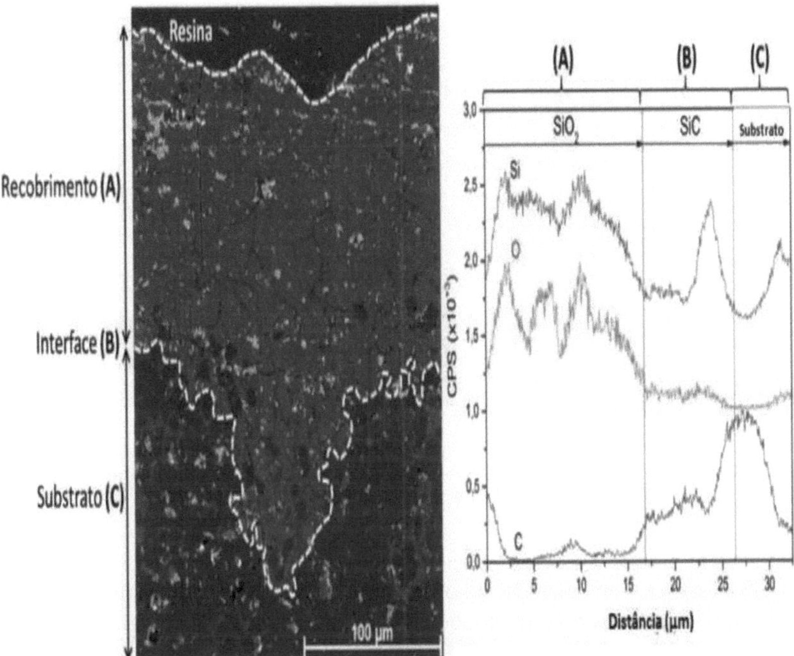

Figure 44. Cross-sectional image of sample SMP 9 and EDS results showing the atomic composition profile of the coating (A), interface (B) and substrate (C).

The region (A) is predominantly composed of Si and O, the region (B) which is the interface between the coating and the substrate shows the presence of Si and C and finally the region (C) which is predominantly composed of C, as it is the substrate. The substrate is obviously a carbon fiber composite with phenolic resin and therefore contains other elements besides carbon.

1.6.2. Nanostructured coatings

Another characteristic of coatings obtained using liquid precursors is the synthesis of nanostructured materials, which have unique characteristics in terms of their mechanical and thermal properties, as well as producing lower defect rates and low porosity when compared to coatings obtained in the traditional way, which use solid raw materials and are limited to the size of the sprayed particle (ZENG, 2002). The nanometric structure of the coating was identified from a crack in the coating. The result shown in Figure 45 was obtained using an SEM/FEG, which allows a higher magnification while maintaining good image resolution. The size of the largest spheres forming the coating were measured and showed diameters ranging from 38~48 nm.

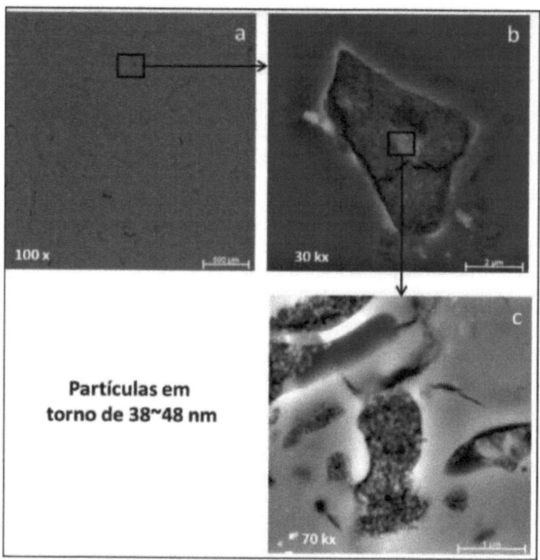

Figure 45. SEM-FEG images showing the formation of nanostructures.

EDS analyses were also carried out at specific points on the coating to check the composition of the nanostructures found. The results are shown in Figure 46.

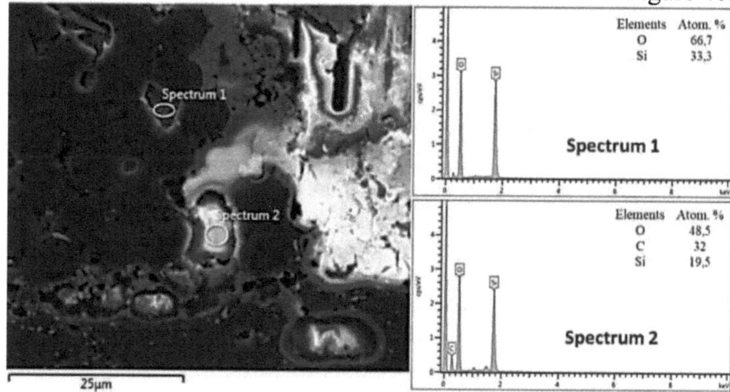

Figure 46. EDS of regions of the nanostructured coating.

The point in the region labeled "*spectrum 1*" is the point where the nanoparticles shown in Figure 45 were found. This region categorically shows the composition of the SiO_2 coating, with a ratio of 1:2 between Si and O. In the other region analyzed, identified as "*spectrum 2*", the presence of C, Si and O is observed, and therefore most likely a region where SiC is formed, as already identified in previous analyses, due to the predominance of carbon in regions close to the substrate.

1.7. Thermal testing of coated samples

In order to verify the thermal stability of the coatings, their microstructural properties and whether the cracks present will not actually compromise the protection function against ablative environments that the substrate should have, i.e. the *self-*

healing capacity of the SiO2 coating, thermal (or ablative) tests were carried out in accordance with ASTM E285-80,
"The standard test method for the oxyacetylene ablation testing of thermal insulation materials" ("Standard Test Method for Oxyacetylene Ablation Testing of Thermal Insulation Materials," 2015). Following this standard, the samples are arranged in a position normal to the axis of the flame promoted by the oxyacetylene torch, controlling the pressures of oxygen and acetylene to ensure a stoichiometric mixture between the two gases. The test begins by igniting the torch at a standard distance of 20 mm from the surface of the substrate. During the tests, the samples are kept in contact with the flame for 90 seconds. The surface temperatures of the samples were continuously monitored by an infrared pyrometer (Thermalert TX Raytek). The maximum surface temperature of the samples during the tests was 1800°C, exceeding the melting temperature of SiO2 at around 1700°C, as can be seen in the phase diagram shown in Figure 47. Under these conditions, the ablation test allows us to assess that if the coating has the ability to self-heal, this ability becomes active.

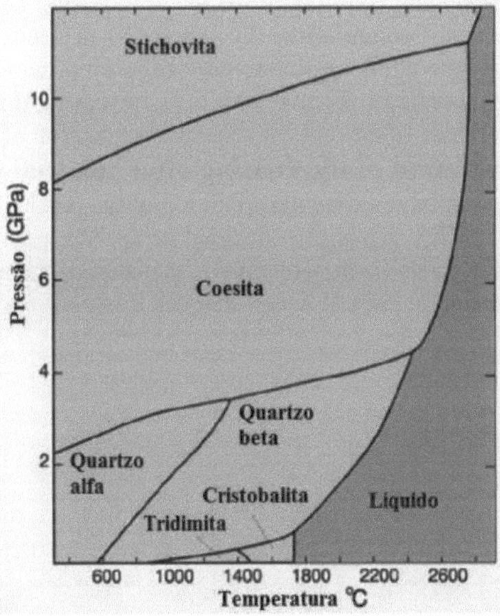

Figure 47. Silicon dioxide phase diagram (FARNESE, 2015)

Figure 48 shows the results of the thermal tests, where Figure 48a refers to the sample before the test and Figure 48b after the ablation test in an ablative oxidizing environment generated by the oxyacetylene torch. It is worth noting that the loss of mass resulting from the tests was not taken into account since the aim was only to validate the self-healing capacity of the coating. This is because the test does not produce a uniform ablation over the entire area of the sample tested, which makes it difficult to assess the specific mass loss rate, i.e. the mass loss as a function of time and area treated.

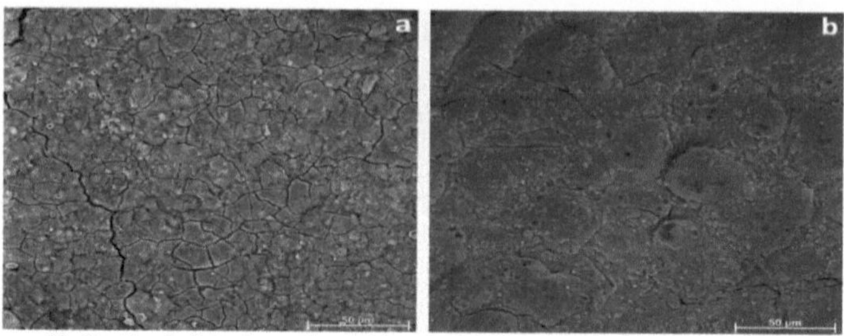

Figure 48. Coating before thermal test (a) and after thermal test (b)

The results show an intense reduction in the cracks present, demonstrating that the coating obtained really does have the ability to self-heal and does not degrade under ablation conditions. For a numerical evaluation of the results, ImageJ software was used to estimate the area of the cracks before and after the ablation test. The cracks present in the sample before the tests corresponded to 8% of the total area of the mapped region, but after the test there was a decrease in the area of the cracks to 1% of the total mapped area. These results prove the self-healing ability of the coatings and that the cracks did not compromise the substrate's protection against oxidizing environments at high temperatures. Another relevant result observed in the images was the maintenance of the coating's thermal stability, preventing direct exposure of the substrate to an oxyacetylene flame and consequent degradation of the composite.

6.7.1 Crystallization of the coating after thermal testing

Since after the thermal tests the densification of the coatings was observed, due to the closure of the cracks, it was also decided to analyze possible phase changes, more precisely, whether crystallization of any of the materials that make up the coating, such as SiO2 and SiC, had occurred. X-ray analysis was used once again. The results are shown in Figure 49.

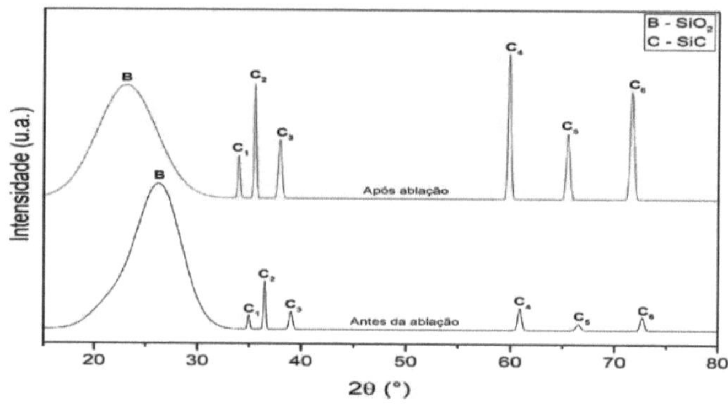

Figure 49. Diffractograms of the sample before the thermal test (-) and after the thermal test (-).

From the diffractogram of the material after the ablation test (red line) it can be seen that the halo referring to amorphous SiO2 decreases in intensity when compared to the material without ablative treatment. Conversely, the diffraction peaks related to SiC increase in intensity, showing the crystallization process of the SiC present in its hexagonal phase, or even new SiC formation reactions.

The hypothesis of new SiC formation reactions during ablation is supported both by observations of the reduction in the intensity of the sio2 halo and by the use of an ablative test environment rich in carbon from the working gas used in the oxyacetylene torch. Due to the high temperature reached and the fact that the SiO2 coating is a nanostructured, amorphous material, it melts at lower temperatures than crystalline material, facilitating the occurrence of the effects analyzed in the ablation tests.

One way of demonstrating that the silicon present in the SiO2 has actually reacted with the carbon is by analyzing the area of the curves corresponding to the peaks of the diffractograms before and after the ablation test (MIRANDA, 2017). The results of the calculations of these areas obtained using the Origin software are listed in Table 4.

Table 4. Demonstration of the areas (in modulus) of the curves corresponding to the peaks observed in the diffractograms shown in Figure 50 and the corresponding differences between the areas before and after the ablative test.

PICOS	AREA 1 (A1)	AREA 2 (A2)	DIFFERENCE *A2-A1*
B	5,90	5,48	0,42
C1	0,02	0,08	0,06
C2	0,08	0,21	0,13
C3	0,04	0,17	0,13
C4	0,06	0,34	0,28
C5	0,02	0,19	0,17
C6	0,04	0,33	0,29

Peaks - Diffraction peaks corresponding to the phases present in the coating.
Area 1- Area of the curves referring to the diffractogram peaks of the samples before the ablation test.
Area 2- Area of the curves referring to the diffractogram peaks of the samples after the ablation test.
Difference *A2-A1* - Difference between the areas of the curves referring to the peaks of the diffractograms of the samples before and after the ablation test (in modulus).

From the results shown in Table 4, it can be seen that the area of the curve relating to the SiO2 (B) peak decreased after the ablation tests were carried out and that all the other peaks relating to hexagonal SiC (C1~C6) saw a gain in their areas, again suggesting that the SiO_2 predominantly in the coating (greater area) reacts with carbon to form a coating "doped" with crystalline SiC.

Final considerations

In this work, a plasma thermal spray system operating in a supersonic regime was developed to study the processes of depositing nanostructured environmental barrier coatings using a precursor based on a colloidal solution. In turn, the coatings are aimed at the environmental protection function of Carbon/Carbon composite substrates of great application in the aerospace sector. The main tool in this system is the non-transferred arc plasma torch, specifically calibrated and characterized to meet the requirements of producing a high enthalpy, high speed plasma jet capable of processing precursors with high melting temperatures. With these characteristics, adjusted to the power, gas and liquid precursor supply systems, nanostructured coatings with low pore fractions and inclusions were obtained. The breakthrough in the method of injecting the material into the torch's mixing chamber was a fundamental step in carrying out the experiments, since the way it works and its construction allowed the experiments to be carried out while inhibiting intense processes of material deposition inside the torch and also in the solution atomizing nozzle, a recurring problem in systems with axial injection of precursors directly into the discharge channel in the region where the electric arc is stabilized. Another advantage of this plasma torch, here called Tandem, is the longer interaction time of the precursor with the high temperature plasma regions, passing through regions adjacent to the arc in the discharge channel, following the torch nozzle and sprayed in the plasma jet region, which also allows the use of precursors in the form of high melting point powders, such as those of ceramic materials. In this context and also on a theoretical level, the study of the heat and mass transport mechanisms of the particles generated in the plasma/precursor interaction process is a complex task, but it is essential for evaluating nanostructure formation processes, since the calculation of the nucleation rate and the condensation rate are important parameters used to determine the particle growth mechanism and the formation of nanoparticles. These mechanisms depend on several factors, including the properties of the plasma gas and the precursor, the power of the torch and the residence time of the particle in plasma regions. In processes using DC or RF plasma torches, the material is typically introduced into the plasma jet in the form of micrometric powders. They are accelerated due to the momentum transfer of the plasma jet and at the same time the heat and mass transfer process begins. The residence time of the powder in the plasma jet, and consequently the efficiency of the process, will depend on the velocity acquired by the particle and the length of the jet. The process of forming nanostructured coatings is related to these factors, since particles inside the plasma flame must remain long enough for them to evaporate completely. The formation of nanostructures occurs with the rapid cooling of the vapor. It is important to note that this process is more commonly observed in works that use short arc plasma torches transferred to make better use of the process energy (PFENDER, 1985; WATANABE, 2009). In these systems, the vapor cooling process is the main stage in the formation of nanoparticles. Homogeneous nucleation is favored by a combination of heating, evaporation and rapid cooling, such as what happens in a plasma reactor (GIRSHICK, 1988). It is important to note that in this work, an atomized solution is injected directly into the discharge channel of the torch, acting in its path similarly to a gas, which will therefore constitute a plasma jet practically free of microparticles with sizes that can be defined, like the one used in the theoretical model presented in this thesis. This model was developed in order to estimate the speed and temperature of microparticles and evaluate possible effects on the properties of coatings.

The fact is that in the present work, nanostructured coatings were obtained using a supersonic non-transferred arc plasma torch and therefore with lower enthalpies than those using a transferred arc. The formation processes of these nanostructures were probably made possible or facilitated by the use of atomized liquid precursors in interaction not only with the plasma jet but also with the arc establishment regions in the discharge channel. The results show the formation of a nanostructured coating made up of particles with a maximum diameter of 47 nm and with defects (cracks and pores) that follow the order of nanometric size.

The process of rapidly cooling the particles on the substrate contributes not only to the synthesis of a nanostructured coating but also an amorphous one, as shown by the results of the microstructural analysis of the coatings obtained. These results correlate with the intrinsic characteristics of SiO2 when exposed to rapid cooling with a consequent reduction in molecular movement, preventing the material from organizing itself to obtain crystalline periodicity (DEBENEDETTI, 2001). However, when considering applications in environmental protection systems, amorphous materials may have advantages over crystalline materials in terms of thermal diffusivity.

Another relevant result obtained during the analysis of the chemical and structural composition of the coating was the formation of SiC due to the interaction processes between the substrate and the liquid precursor plasma jet. The formation of SiC in the coating helps to protect the composite, as it reduces oxygen permeability, preventing its oxidation and consequently the loss of its structural characteristics. Although no coating adhesion tests have been carried out, the higher concentration of SiC at the coating-composite interface indicates the occurrence of a coating chemical adhesion process (STERN, 1996). Thus, at this stage of the deposition process, the composite is "doped" with a SiC layer and is therefore exposed to different oxidation processes compared to the original C/C substrate. In particular, the SiC oxidation mechanisms can be passive or active. Although there are additional decomposition reactions or mechanisms, the most likely chemical oxidation reactions for SiC are as follows (HALD, 2003):

Passive Oxidation
Active Oxidation

$$2SiC(s)+3O_2(g) \rightarrow 2SiO_2(s,l)+2CO(g)$$

$$SiC(s)+O_2(g) \rightarrow SiO(g)+CO(g)$$

$$SiC(s)+SiO_2(s,l) \rightarrow 3SiO(g)+CO(g)$$

In situations of high temperature and low oxidizer partial pressure, the formation of a protective SiO2 layer on the surface of the composite is suppressed and only volatile CO and SiO gases are formed, leading to rapid consumption of the SiC and

SiO2 present in the composite. This behavior is called "active oxidation", which should occur in our case when the samples are closest to the plasma torch nozzle. The other oxidation mechanism, called "passive", occurs preferentially at lower temperatures and higher partial pressures of oxidizer. In this regime, a stable $SiO_2(l)$ layer forms on the surface, which reduces the diffusion of oxidizers into the composite and consequently also the rate of mass loss. On the other hand, the CO gas generated at this interface flows by diffusion and/or permeation in the opposite direction.

In the reverse route, SiO_2 is distributed, with a higher concentration on the surface of the coating exposed to the environment. For the thicker coatings synthesized, the presence of SiC on the surface was not detected, indicating that the formation of a SiO2 passivation layer on the surface of the coating is intensified with deposition time. This layer not only inhibits the diffusion of oxygen and nitrogen (generated in the plasma jet) towards the substrate, maintaining its integrity, but also inhibits the diffusion of carbon from the substrate to the coating surface, reducing the concentration of SiC in this region.

The effect of the flow velocity of reactive particles on the oxidation mechanisms of SiC is rarely reported in the literature. SiO2 under attack by erosive oxygen reactions, depending on the speed of the oxygen flow, can lead to the formation of gaseous SiO or can only lead to exchange reactions that do not alter the material's mass balance. In most cases, both types of reactions occur simultaneously (DABALÀ, 1995). Tests carried out at low pressure
110 (1000 Pa) in a plasma tunnel showed that the increase in flow velocity and pressure leads to a strong increase in exchange reactions due to the high transport rates that occur in high velocity plasmas. According to Dabala and colleagues (DABALÁ, 1995), the increase in the flow of oxygen to the reaction site due to the increase in flow velocity is not comparable to the effect of a higher partial pressure of oxygen, but can lead to the transport of volatile intermediate products away from the reaction site. This type of mechanism is capable of breaking down an existing SiO2 layer or even preventing its formation. This is important not only for the passive-active transition of reaction conditions, but depending on the flow velocity it can also impair the formation of the protective layer under passivation conditions. According to theoretical models simulated at pressures of 1 bar (DABALÁ, 1995), the SiO2 sublimation mechanism is strengthened by the increase in flow velocities and also by the high temperature of the exposed surface, which can therefore initiate the transition to active oxidation of the SiC layer formed close to the substrate by exposing it directly to the plasma jet. This effect is probably predominant in the case of static samples or those very close to the torch nozzle where, in the present work, intense coating degradation was observed.

With regard to the effect of the interaction of high kinetic energy atomic oxygen in regions with high concentrations of C, Si and SiC, (making an equivalence with regard to the material formed at the substrate-coating interface), *Electron Probe Micro-Analyzer (*EPMA) analyses (FUJIMOTO, 2003) show that a small amount of oxygen is implanted in regions containing only Si and SiC and almost no oxygen is implanted in the C/C composite region. According to Fugimoto et. al (FUJIMOTO, 2003), the amount of oxygen implanted in the Si and SiC regions increases with exposure time. It is therefore expected that in the C/C composite region, the carbon atoms react with the high energy atomic oxygen and are ejected from the surface in the form of CO, while in the Si or SiC region, the atomic oxygen with high kinetic energy is only scattered or implanted on the surface, without reacting. It can also be concluded from these results that the carbon-only region of the composite is more catalytic or more prone to exothermic recombination reactions with atomic species than the regions containing Si, SiC and SiO_2.

Finally, specific modifications were made to the plasma spray system to operate with liquid precursors in order to obtain nanostructured environmental barrier coatings, which therefore have different characteristics from those obtained using the conventional plasma spray technique. Another important factor is the lower cost of obtaining the precursor using a process also developed by the group. By exploiting the properties of the plasma jet combined with the characteristics of the system used and the substrate, as presented in this thesis, a coating with a compositional gradient between SiC and SiO_2 was obtained. Reactions that form SiC are rarely reported in the literature, especially at atmospheric pressure and using conventional spray plasma systems. The study carried out and presented in this paper expands the system's operating capacity to deposition processes, indicating potential also in the manufacture of other nanostructured materials and nanoparticles.

Conclusions

The plasma thermal spray deposition process and system using liquid precursors were developed. Energy characterizations of the plasma torch were carried out to define the operating parameters used during the depositions. The technique developed made it possible to obtain coatings with a compositional gradient between SiO_2/SiC that adhered to the C/C composite used as a substrate.

The formation of SiC under atmospheric pressure is an innovative result of the process. According to the results of the chemical analysis, the SiC is more concentrated at the substrate/coating interface, made possible by reactions between the Si in the precursor and the carbon in the substrate, thus giving the coating chemical adhesion without the need to synthesize a *bond coating*.

The FTIR and Raman analyses evaluated the influence of the deposition time on the chemical composition of the coating, by means of the ratio between the areas of the SiO_2 and SiC peaks.

Coatings were obtained at an average deposition rate of 6.4 LinVmin, with a predominance of SiO_2 formation reactions on the surface of the coating with increasing exposure time and, consequently, the thickness of the deposited layer.

The microstructural characteristics of the coating were evaluated using SEM-FEG images, which showed the formation of cracks after cooling. However, these cracks did not compromise the integrity of the coatings due to the self-healing ability of silica when exposed to thermal tests in oxidizing environments. The coatings obtained had a nanometric structure with an average particle size distribution of approximately 39 nm, which has a direct impact on the porosity, thermal conductivity, deformation tolerance and elastic modulus of the material.

On a theoretical level, the results obtained helped to discuss and understand the behavior of the particles in relation to their speed and temperature. The increase in the initial velocity of the particle, with a plasma jet with fixed parameters (such as constant temperature and velocity), determines the total energy of the particles. In general, increasing the kinetic energy of the particle results in lower thermal energy. For the model used, the particle reaches the highest total energy (kinetic and thermal) only when the injection velocity is equal to zero. As a result, although it is not the case in this thesis, the impact of microparticles at high speed and temperature contributes to better adhesion of the coating, as well as reducing defects and porosity.

In relation to the structure of the coatings, the X-ray diffractograms showed that the coating is amorphous, a result directly related to the characteristics of the process, which provides high heat transfer between plasma and particles, causing the particles to be heated and cooled rapidly, inhibiting crystallization of the material.

The results obtained correlate directly with the innovative characteristics of the system developed, demonstrating the great potential for application of the HVSPS deposition technique.

Suggestions for future work

The successful development of the system for processing liquid precursors has led to the possibility of new research in the area of materials deposition and fabrication. Highlights include:

- Theoretical studies (simulations) of the behavior and formation of plasma torch-processed materials from liquid precursors, including studies of the formation and nucleation of nanoparticles;
- Ablation tests of the materials obtained and of new materials are carried out in a supersonic plasma tunnel that simulates the atmospheric re-entry environment;
- Testing the adhesion of the coating to the substrate;
- Production of nanoparticles using liquid routes for high added value products;
- Processing of ultra-high-temperature ceramics (UHTC);
- Evaluation of the effect of plasma jet parameters (temperature, speed, enthalpy) on coating properties (deposition rate, composition, porosity, adhesion, etc.)

References

ABOURIDA, M. AND HARB, F. Synthesis and Characterization of Amorphous Silica Nanoparticles from Aqueous Silicates Using Cationic Sulfactants. **Journal of Metals, Materials and Minerals,** v. 24, n. 1, p. 37-42, 2014.

ALFANO, D.; SCATTEIA, L.; CANTONI, S.; BALAT-PICHELIN, M. Emissivity and catalycity measurements on SiC-coated carbon fiber reinforced silicon carbide composite. **Journal of the European Ceramic Society**, v. 29, n. 10, p. 2045-2051, 2009a.

ALFANO, D.; SCATTEIA, L.; CANTONI, S.; BALAT-PICHELIN, M. Emissivity and catalycity measurements on SiC-coated carbon fiber reinforced silicon carbide composite. **Journal of the European Ceramic Society**, v. 29, n. 10, p. 2045-2051, 2009b.

ALLAIRE, F.; PARENT, L.; DALLAIRE, S. Production of submicron SiC particles by d.c. thermal plasma: a systematic approach based on injection parameters.
Journal of Materials Science, v. 26, n. 15, p. 4160-4165, 1991.

ANDO, K.; FURUSAWA, K.; TAKAHASHI, K.; SATO, S. Crack-healing ability of structural ceramics and a new methodology to guarantee the structural integrity using the ability and proof-test. **Journal of the European Ceramic Society**, v. 25, n. 5, p. 549-558, 2005.

BANSAL, N. P.; DOREMUS, R. H. Handbook of Glass Properties. Handbook of **Glass Properties**, p. 1-680, 2013.

BARBOUCHE, M.; ZAGHOUANI, R. B.; BENAMMAR, N. E.; KHIROUNI, K.
Synthesis and characterization of 3C-SiC by rapid silica carbothermal reduction. **The International Journal of Advanced Manufacturing Technology**, p. 1339-1345, 2017.

BELMONTE, M.; GONZALEZ-JULIAN, J.; MIRANZO, P.; OSENDI, M. I. Continuous in situ functionally graded silicon nitride materials. **Acta Materialia**, v. 57, n. 9, p. 2607-2612, 2009.

BERTOLUZZA, A.; FAGNANO, C.; ANTONIETTA MORELLI, M.; GOTTARDI, V.; GUGLIELMI, M. Raman and infrared spectra on silica gel evolving towards glass. **Journal of Non-Crystalline Solids**, v. 48, n. 1, p. 117-128, 1982.

BOULOS, M. I.; FAUCHAIS, P.; PFENDER, E. **Thermal Plasma Fundamentals and Applications**. 1st ed. New York: Springer Science+Business Media, LLC, 1994.

BUCKLEY, J. D.; EDIE, D. D. **Carbon-carbon Materials and Composites**. 1st ed. New Jersey: Noyes Publications, 1993.

CALIARI, F. R. **Development of thermal plasma deposition process of ceramic and metallic materials on titanium alloy substrate for aerospace applications**, 2016. 119 f.Thesis (PhD Materials and Processes for Industrial Applications) Universidade Federal de Sao Paulo, Sao José dos Campos.

CAO, X. Q.; VASSEN, R.; STOEVER, D. Ceramic materials for thermal barrier coatings. **Journal of the European Ceramic Society**, v. 24, n. 1, p. 1-10, 2004. CHENG, L.; XU, Y.; ZHANG, L.; YIN, X. Oxidation behavior of three dimensional C/SiC composites in air and combustion gas environments. **Carbon**, v. 38, p. 21032108, 2000.

COUNCIL, N. R. **Coatings for High-Temperature Structural Materials: Trends and Opportunities**. Washington, DC: The National Academies Press, 1996.

DABALÀ, P.; HILFER, G.; AUWETER-KURTZ, M. Investigation of the oxidation behavior of thermal protection materials supported by mass spectrometry.
Aerothermodynamics for space vehicles, of the 2nd European Symposium, Noordwijk , 1995.

DAI, X.; YAN, D.; YANG, Y.; et al. In situ (Al , Cr) 2 O 3 -Cr composite coating fabricated by reactive plasma spraying. **Ceramics International**, v. 43, n. 8, p. 63406344, 2017.

DAVIS, J. R. **Handbook of Thermal Spray Technology**. 1ed. ASM International. 2004.

DEBENEDETTI, P. G.; STILLINGER, F. H. Full-Text. **Nature**, v. 410, n. March, p. 259-267, 2001.

DECOTTIGNIES, M.; PHALIPPOU, J.; ZARZYCKI, J. Synthesis of glasses by hot- pressing of gels. **Journal of Materials Science**, v. 13, n. 12, p. 2605-2618, 1978.

DELBOS, C.; FAZILLEAU, J.; RAT, V.; et al. Phenomena involved in suspension plasma spraying part 2: Zirconia particle treatment and coating formation. **Plasma Chemistry and Plasma Processing**, v. 26, n. 4, p. 393-414, 2006.

ENGEL, T.; KICKELBICK, G. Self-healing nanocomposites from silica - polymer core - shell nanoparticles. **Polymer International**, v. 63, n. 5, p. 915-923, 2014.

ESSIPTCHOUK, A. M.; CHARAKHOVSKI, L. I.; FILHO, G. P.; et al. Thermal and power characteristics of plasma torch with reverse vortex. **Journal of Physics D: Applied Physics**, v. 42, n. 17, p. 175205, 2009.

ESSIPTCHOUK, A.; PETRACONI, G.; CALIARI, F. R.; et al. Numerical Study of Particle Heating in a Plasma Jet. **Journal of Engineering Physics and Thermophysics**, v. 90, n. 2, 2017.

FAHRENHOLTZ, W. G. The ZrB 2 volatility diagram. **Journal of the American Ceramic Society**, v. 88, n. 12, p. 3509-3512, 2005.

FAHRNER, W. R. **Amorphous silicon / crystalline silicon heterojunction solar cells**. 1st ed. Springer Science+Business Media, LLC, 2013.

FAN, W.; BAI, Y. Review of suspension and solution precursor plasma sprayed thermal barrier coatings. **Ceramics International**, 2016.

FARNESE, C. **Deposition of Silicon Dioxide Film on C/C Composite by**

Plasma Spray from Colloidal Solution, 2015. 86f. Dissertation (Master's Degree in Aeronautical and Mechanical Engineering), Instituto Tecnológico de Aeronáutica, Sao José dos Campos.

FAUCHAIS, P. Understanding plasma spraying. **Journal of Physics D: Applied Physics**, v. 37, n. 9, p. R86-R108, 2004.

FAUCHAIS, P.; ETCHART-SALAS, R.; RAT, V.; et al. Parameters controlling liquid plasma spraying: Solutions, sols, or suspensions. **Journal of Thermal Spray Technology**, v. 17, n. 1, p. 31-59, 2008.

FAUCHAIS, P. L.; HEBERLEIN, J. V. R.; BOULOS, M. I. **Thermal Spray Fundamentals**. 2014.

FAUCHAIS, P.; RAT, V.; COUDERT, J. F.; ETCHART-SALAS, R.; MONTAVON, G. Operating parameters for suspension and solution plasma-spray coatings. **Surface and Coatings Technology**, v. 202, n. 18, p. 4309-4317, 2008.

FAUCHAIS, P.; VARDELLE, M.; GOUTIER, S.; VARDELLE, A. Key Challenges and Opportunities in Suspension and Solution Plasma Spraying. **Plasma Chem Plasma Process**, v. 35, p. 511-525, 2015.

FORD, R. G. **Functionally Graded Materials: Design, Processing and Applications**. 1st ed. New York: Springer Science+Business Media, LLC, 1999.

FUJIMOTO, K. Degradation of carbon-based materials due to impact of high-energy atomic oxygen. **International Journal of Impact Engineering**, v. 28, n. 1, p. 1-11, 2003.

GELL, M. Application opportunities for nanostructured materials and coatings. **Materials Science and Engineering A**, v. 204, n. 1-2, p. 246-251, 1995.

GELL, M.; JORDAN, E. H.; TEICHOLZ, M.; et al. Thermal Barrier Coatings Made by the Solution Precursor Plasma Spray Process. **Journal of Thermal Spray Technology**, v. 17, n. March, p. 124-135, 2008.

GHOSH, S. Thermal Barrier Ceramic Coatings - A Review. **Advanced Ceramic Processing**. p.135-152, 2015. InTech.

GHOSH, S. K. Self-Healing Materials: Fundamentals, Design Strategies, and Applications. **Self-Healing Materials**. p.1-28, 2008.

GIRSHICK, S. L.; CHIU, C. P.; MCMURRY, P. H. Modelling particle formation and growth in a plasma synthesis reactor. **Plasma Chemistry and Plasma Processing**, v. 8, n. 2, p. 145-157, 1988.

GUNDIAH, G.; MADHAV, G. V.; GOVINDARAJ, A.; SEIKH, M. M.; RAO, C. N. R. Synthesis and characterization of silicon carbide, silicon oxynitride and silicon nitride nanowires. **Journal of Materials Chemistry**, v. 12, n. 5, p. 1606-1611, 2002.

GUNDIAH, G.; MADHAV, G. V; GOVINDARAJ, A.; SEIKH, M. M.; RAO, C. N. R. Synthesis and characterization of silicon carbide, silicon oxynitride and silicon nitride nanowires. **Journal of Materials Chemistry**, v. 12, n. 5, p. 1606-1611, 2002.

GUO, F. A.; TRANNOY, N.; GERDAY, D. An application of scanning thermal microscopy: Analysis of the thermal properties of plasma-sprayed yttria-stabilized zirconia thermal barrier coating. **Journal of the European Ceramic Society**, v. 25, n. 7, p. 1159-1166, 2005.

GUO, Y. P.; ZHENG, J. C.; WEE, A. T. S.; et al. Photoluminescence studies of SiC nanocrystals embedded in a SiO 2 matrix. **Chemical Physics Letters**, v. 339, n. May, p. 319-322, 2001.

HALD, H. Operational limits for reusable space transportation systems due to physical boundaries of C/SiC materials. **Aerospace Science and Technology**, v. 7, n.
7, p. 551-559, 2003.

HERMANEK, F. J. **Thermal Spray Terminology and Company Origins**. ASM International , p. 73, 2001.

HU, P.; GUOLIN, W.; WANG, Z. Oxidation mechanism and resistance of ZrB2-SiC composites. **Corrosion Science**, v. 51, n. 11, p. 2724-2732, 2009.

HUANG, J.; LI, H.; ZENG, X.; LI, K. Yttrium silicate oxidation protective coating for SiC coated carbon / carbon composites. **Ceramics International** , v. 32, p. 417421, 2006.

HUB, S. L. **Plasma spray-coating**. Available at: <http://legacy.sciencelearn.org.nz/Contexts/Gases-and-Plasmas/Sci-Media/Images/Plasma-spray-process>. Accessed on: 10 Dec.2017.

HUNT, A. M. W. (ED.). **The Oxford Handbook of Archaeological Ceramic Analysis**. Oxford University Press, 2016.

HWANG, S. S.; VASILIEV, A. L.; PADTURE, N. P. Improved processing and oxidation-resistance of ZrB2 ultra-high temperature ceramics containing SiC nanodispersoids. **Materials Science and Engineering A**, v. 464, n. 1-2, p. 216-224, 2007.

KAISER, A.; LOBERT, M.; TELLE, R. Thermal stability of zircon (ZrSiO 4). **Journal of the European Ceramic Society**, v. 28, p. 2199-2211, 2008.

KHOR, K. A.; DONG, Z. L.; GU, Y. W. Plasma sprayed functionally graded thermal barrier coatings. **Materials Letters**, v. 38, n. 6, p. 437-444, 1999.

KIEBACK, B.; NEUBRAND, A.; RIEDEL, H. Processing techniques for functionally graded materials. **Materials Science and Engineering A**, v. 362, n. 1-2, p. 81-105, 2003.

KIM, J. I.; KIM, W. J.; CHOI, D. J.; PARK, J. Y.; RYU, W. S. Design of a C/SiC functionally graded coating for the oxidation protection of C/C composites. **Carbon**, v. 43, n. 8, p. 1749-1757, 2005.

KOKINI, K.; CHOULES, B. D.; TAKEUCHI, Y. R. Thermal fracture mechanisms in ceramic thermal barrier coatings. **Journal of Thermal Spray Technology**, v. 6, n. 1, p. 43-49, 1997.

KOVALEV, V. L.; KOLESNIKOV, A. F. Experimental and theoretical simulation of heterogeneous catalysis in aerothermochemistry (a review). **Fluid Dynamics**, v. 40, n. 5, p. 669-693, 2005.

KULKARNI, A.; VAIDYA, A.; GOLAND, A.; SAMPATH, S.; HERMAN, H.
Processing effects on porosity-property correlations in plasma sprayed yttria- stabilized zirconia coatings. **Materials Science and Engineering: A**
, v. 359, p. 100111, 2003.

LANGE, F.F. AND RADFORD, K. C. Healing of surface cracks in polycrystalline Al2O3. **Journal of the American Ceramic Society**. v. 53, n. 7, 1970.

LEE, K. N. Current status of environmental barrier coatings for Si-based ceramics. **Surface and Coatings Technology**, v. 133-134, p. 1-7, 2000.

LEE, K. N. U. Current status of environmental barrier coatings for Si-Based ceramics. **Surface and coatings technology**, p. 1-7, 2000.

LEVY NETO, F.; PARDINI, L. C. LEVY NETO, F. and PARDINI, L. C. **estrutu rais ciência e tecno lo gia** . Sâo Paulo: Edgar Blücher, 2006. p. 336. Sâo Paulo: Edgar Blücher, 2006.

LI, N.; PRONK, A.; MOLENAAR, A.; VEN, M. VAN DE; WU, S. Comparison of Uniaxial and Four-Point Bending Fatigue Tests for Asphalt Mixtures.
Transportation Research Record: Journal of the Transportation Research Board, v. 2373, p. 44-53, 2013.

LIU, J.; CAO, L.; HUANG, J.; et al. Surface & Coatings Technology A ZrSiO 4 / SiC oxidation protective coating for carbon / carbon composites. **Surface & Coatings Technology**, v. 206, n. 14, p. 3270-3274, 2012. Elsevier B.V.

MARQUESI, A. R. **Study and development of a water vapor plasma torch for applications in gasification processes**, 2016. 158 f. Thesis (Doctorate in Plasma Physics) - Instituto Tecnológico de Aeronáutica, Sâo José dos Campos.

MI, P.; HE, J.; QIN, Y.; CHEN, K. Nanostructure reactive plasma sprayed TiCN coating. **Surface and Coatings Technology**, v. 309, p. 1-5, 2017.

MIRANDA, F. S.; CALIARI, F. R.; CAMPOS, T. M.; ESSIPTCHOUK, A. M.;
FILHO, G. P. Deposition of graded SiO2/ SiC coatings using high-velocity solution plasma spray. **Ceramics International**, v. 43, n. 18, p. 16416-16423, 2017.

MOON, S.; HATANO, M.; LEE, M.; GRIGOROPOULOS, C. P. Thermal conductivity of amorphous silicon thin films. **International Journal of Heat and
Mass Transfer**, v. 45, n. 12, p. 2439-2447, 2002.

PALANINATHAN, R. Behavior of Carbon-Carbon Composite under Intense Heating. **International Journal of Aerospace Engineering**, v. 2010, n. i, p. 1-7, 2010.

PALANINATHAN, R.; BINDU, S. Modeling of Mechanical Ablation in Thermal Protection Systems. **Journal of Spacecraft and Rockets** , v. 42, n. 6, 2005.

PAWLOWSKI, L. **The Science and Engineering of Thermal Spray Coatings**.
Second ed. John Wiley & Sons, Ltd, 2007.
PAWLOWSKI, L. Finely grained nanometric and submicrometric coatings by thermal spraying: A review. **Surface and Coatings Technology**, v. 202, n. 18, p. 4318-4328, 2008.
PAWLOWSKI, L. Suspension and solution thermal spray coatings. **Surface and Coatings Technology**, v. 203, n. 19, p. 2807-2829, 2009.
PFENDER, E.; LEE, Y. C. Particle dynamics and particle heat and mass transfer in thermal plasmas. Part I. The motion of a single particle without thermal effects. **Plasma Chemistry and Plasma Processing**, v. 5, n. 3, p. 211-237, 1985.
QI, X.; ZHAI, G.; LIANG, J.; et al. Preparation and characterization of SiC@CNT coaxial nanocables using CNTs as a template. **CrystEngComm**, v. 16, n. 41, p. 9697-9703, 2014. Royal Society of Chemistry.
QUINN, T. P. Second Edition, Revised and Expanded. **The Quarterly Review of Biology**, v. 80, n. 1, p. 128-129, 2005.
SAMPATH, S.; JIANG, X. Splat formation and microstructure development during plasma spraying: deposition temperature effects. **Materials Science and Engineering: A**, v. 304-306, n. 1-2, p. 144-150, 2001.
SARKISOV, P. D.; POPOVICH, N. V.; ORLOVA, L. A.; ANAN&APOS;EVA, Y. E. Barrier coatings for type C/SiC ceramic-matrix composites (Review). **Glass and Ceramics (English translation of Steklo i Keramika)**, v. 65, n. 9-10, p. 366-371, 2008.
SAVAGE, G. **Carbon-Carbon Composites**. 1st ed. Springer Science+Business Media, LLC, 1993.
SCHULZ, U.; PETERS, M.; BACH, F. W.; TEGEDER, G. Graded coatings for thermal, wear and corrosion barriers. **Materials Science and Engineering A**, v. 362, n. 1-2, p. 61-80, 2003.
SILVA, G, W. **Qualification of Materials Used in Thermal Protection Systems for Space Vehicles**. 2009. 112f Dissertation (Master's Degree), Technological Institute of Aeronautics, Sao José dos Campos.
SILVA, R. J. **Thermal plasma for ablation of materials used as thermal protection shields in aerospace systems**, 2011. 132f. Dissertation (Master's Degree) - Technological Institute of Aeronautics, Sao José dos Campos.
SINGH, H.; SIDHU, T. S.; KALSI, S. B. S. Cold spray technology: Future of coating deposition processes. **Frattura ed Integrita Strutturale**, v. 22, p. 69-84, 2012.
SOLONENKO, O. Thermal Plasma Torches and Technologies: Plasma Torches, Basic Studies and Design. **Annals of Physics**, p. 397, 2003.
ASTM International. **E285-08**: Standard Test Method for Oxyacetylene Ablation Testing of Thermal Insulation Materials.West Conshohocken,

2015.6p.

STERN, K. H. **Metallurgical and Ceramic Protective Coatings**. 1st ed. Washington, DC: Chapman & Hall, 1996.

TKACHENKO, L. A.; SHAULOV, A. Y.; BERLIN, A. A. High-temperature protective coatings for carbon fibers. **Inorganic Materials**, v. 48, n. 3, p. 213-221, 2012.

TUCKER, R. C.; SOCIETY, A. S. M. I. T. S.; COMMITTEE, A. S. M. I. H. **ASM Handbook: Thermal spray technology. Volume 5A**. ASM International, 2013.

TULUI, M.; LIONETTI, S.; PULCI, G.; et al. Zirconium diboride based coatings for thermal protection of re entry vehicles: Effect of MoSi2 addition. **Surface and Coatings Technology**, v. 205, n. 4, p. 1065-1069, 2010.

VARDELLE, A.; MOREAU, C.; THEMELIS, N. J.; CHAZELAS, C. A Perspective on Plasma Spray Technology. **Plasma Chemistry and Plasma Processing**, p. 491509, 2014.

VISWANATHAN, V.; LAHA, T.; BALANI, K.; AGARWAL, A.; SEAL, S. Challenges and advances in nanocomposite processing techniques. **Materials Science and Engineering R: Reports**, v. 54, n. 5-6, p. 121-285, 2006.

WATANABE, T.; TANAKA, M. Thermal Plasma Processing for Functional Nanoparticle Synthesis. **16th Asean regional Symposium on Chemical Engineering**, , n. 1, p. 4, 2009.

WEBSTER, J. D.; DAY, R. J.; HAYES, F. H.; et al. Oxidation protection for carbon fiber composites. **Journal of Materials Science**, v. 31, p. 1389-1397, 1996.

WOOL, R. P. Self-healing materials : a review. **The Royal Society of Chemistry**, p. 400-418, 2008a.

WOOL, R. P. Self-healing materials: a review. **Soft Matter**, v. 4, n. 3, p. 400, 2008b.

WRIGHT, W. W. **International encyclopedia of composites: Volumes 1 & 2** Edited by Stuart M. Lee, VCH Publishers, New York, 1990.

XIE, L.; CHEN, D.; JORDAN, E. H.; et al. Formation of vertical cracks in solutionprecursor plasma-sprayed thermal barrier coatings. **Surface and Coatings Technology**, v. 201, n. 3-4, p. 1058-1064, 2006.

XIE, L.; MA, X.; OZTURK, A.; et al. Processing parameter effects on solution precursor plasma spray process spray patterns. **Surface and Coatings Technology**, v. 183, n. 1, p. 51-61, 2004.

YANG, Y.; LI, K.; ZHAO, Z.; LI, H. Ablation resistance of HfC-SiC coating prepared by supersonic atmospheric plasma spraying for SiC-coated C/C composites. **Ceramics International**, v. 42, n. 4, p. 4768-4774, 2016.

ZENG, Y.; LEE, S. W.; GAO, L.; DING, C. X. Atmospheric plasma sprayed coatings of nanostructured zirconia. **Journal of the European**

Ceramic Society, v. 22, p. 347-351, 2002.

ZHUKOV, M. F.; ZASYPKIN, I. M.; TIMOSHEVSKII, A. N.; MIKHAILOV, B. I.; DESYATKOV, G. A. **Thermal Plasma Torches: Design, Characteristics, Application**. Cambridge: CAMBRIDGE INTERNATIONAL SCIENCE PUBLISHING, 2007.

ZWAAG, S. VAN DER. Routes and mechanisms towards self healing behavior in engineering materials. **Bulletin of the Polish Academy of Sciences: Technical Sciences**, v. 58, n. 2, p. 227-236, 2010.